Luminos is the Open Access monograph publishing program
from UC Press. Luminos provides a framework for preserving and
reinvigorating monograph publishing for the future and increases
the reach and visibility of important scholarly work. Titles published
in the UC Press Luminos model are published with the same high
standards for selection, peer review, production, and marketing as
those in our traditional program. www.luminosoa.org

The publisher and the University of California Press Foundation gratefully acknowledge the generous support of the Fletcher Jones Foundation Imprint in Humanities.

Music Streaming around the World

Music Streaming around the World

Edited by
David Hesmondhalgh

UNIVERSITY OF CALIFORNIA PRESS

University of California Press
Oakland, California

Suggested citation: Hesmondhalgh, D., ed. *Music Streaming around the World*. Oakland: University of California Press, 2025. DOI: https://doi.org/10.1525/luminos.248

Library of Congress Cataloging-in-Publication Data

Names: Hesmondhalgh, David, editor.
Title: Music streaming around the world / edited by David Hesmondhalgh.
Description: Oakland, California : University of California Press, [2025] | Includes bibliographical references and index.
Identifiers: LCCN 2025004376 (print) | LCCN 2025004377 (ebook) | ISBN 9780520422599 (cloth) | ISBN 9780520409057 (paperback) | ISBN 9780520409071 (ebook)
Subjects: LCSH: Streaming audio. | Music and the Internet. | Sound recording industry. | Music and globalization.
Classification: LCC ML74.7 .M84 2025 (print) | LCC ML74.7 (ebook) | DDC 780.285/4678—dc23/eng/20250305

LC record available at https://lccn.loc.gov/2025004376
LC ebook record available at https://lccn.loc.gov/2025004377

GPSR Authorized Representative: Easy Access System Europe, Mustamäe tee 50, 10621 Tallinn, Estonia, gpsr.requests@easproject.com

34 33 32 31 30 29 28 27 26 25
10 9 8 7 6 5 4 3 2 1

CONTENTS

ACKNOWLEDGMENTS

This book emerged from a symposium held at the University of Leeds in January 2023, organized as part of MUSICSTREAM: Music Culture in the Age of Streaming, a research project funded by an Advanced Research Grant awarded by the European Research Council, under the European Union's Horizon 2020 Research and Innovation Programme (grant agreement no. 1010020615). The MUSICSTREAM team consists of me as principal investigator, my coinvestigator Shuwen Qu of Jinan University (Guangzhou), and postdoctoral researchers Raquel Campos Valverde, D. Bondy Valdovinos Kaye, and Zhongwei (Mabu) Li. The chapters published under our names are based on research funded by the same grant. The grant has also paid for open-access publication of all the chapters in the book.

For her chapter, Darci Sprengel received funding from the European Research Council (ERC) under the European Union's Horizon 2020 research and innovation programme: grant agreement No 101019164), Principal Investigator Prof. Georgina Born, host institution University College London.

Zoe Cosker provided great assistance in organizing the symposium and getting the project under way before she departed to begin her own research career. Zoe, you're missed!

I'm extremely grateful to the team, all the contributors to this volume, and Raina Polivka, Sam Warren, Stephanie Summerhays, and Nicholas Taylor at the University of California Press. There are so many other people I could thank here, for dialogue and conversation over the years about culture, politics, music, digital this, and platform that, and for friendship, support, and love. Listing them would take pages, so I hope it's not offensive if I confine myself to thanking my wife, Helen Steward; our wonderful children, Rosa and Joe; my extraordinary sister, Julie; my friend Graham Caveney; and my mum, Maureen (or Mo).

David Hesmondhalgh
Ilkley, England, April 2025

CONTRIBUTORS

EMÍLIA BARNA is Associate Professor at the Department of Sociology and Communication, Faculty of Economic and Social Sciences, Budapest University of Technology and Economics (Hungary).

RAQUEL CAMPOS VALVERDE is a Postdoctoral Research Fellow in the School of Media and Communication at the University of Leeds (UK).

YIU FAI CHOW is Professor at the School of Arts and Social Sciences, Hong Kong Metropolitan University (China). He is also an award-winning writer in pop lyrics and prose.

FRANCESCO D'AMATO is Associate Professor of Cultural Production and Music Studies in the Department of Communication and Social Research at the Sapienza Università di Roma (Italy).

ANDREW J. EISENBERG is Associate Professor of Music at New York University Abu Dhabi (UAE) and Global Network Associate Professor of Music at New York University (USA).

IGNACIO GALLEGO is Lecturer in the Department of Communication at the Universidad Carlos III de Madrid (Spain).

RODRIGO GÓMEZ is Professor of Communications Industries and Policies at the Universidad Autónoma Metropolitana-Cuajimalpa (Mexico) and María Zambrano Fellow at the Universidad Carlos III de Madrid (Spain).

DAVID HESMONDHALGH is Professor of Media, Music and Culture in the School of Media and Communication at the University of Leeds (UK).

D. BONDY VALDOVINOS KAYE is a Postdoctoral Research Fellow in the School of Media and Communication at the University of Leeds (UK).

ADITYA LAL is a PhD student in the People, Work and Employment Department at Leeds University Business School and is also affiliated with the School of Media and Communication at the University of Leeds (UK). He formerly worked in the music industry in India.

ZHONGWEI (MABU) LI is a Postdoctoral Research Fellow in the School of Media and Communication at the University of Leeds (UK).

NORIKO MANABE is Professor and Chair of Music Theory at the Jacobs School of Music, Indiana University Bloomington (USA), with affiliations in the departments of Folklore and Ethnomusicology and East Asian Languages and Cultures.

JEREMY WADE MORRIS is Professor of Media and Cultural Studies in the Department of Communication Arts at the University of Wisconsin–Madison (USA).

ARGELIA MUÑOZ-LARROA is Associate Researcher at the Center for Research in North America (CISAN), Universidad Nacional Autónoma de México (UNAM).

SHUWEN QU is Associate Professor of Music and Media in the School of Journalism and Communication at Jinan University (Guangzhou, China).

ONUR SESIGÜR is Assistant Professor of New Media in the Faculty of Communication at Kadir Has University (Turkey). He formerly worked in the music industry in Turkey.

DARCI SPRENGEL is a Lecturer in the Music Industry in the Department of Culture, Media and the Creative Industries at King's College London (UK).

1

The Global Spread of Music Streaming

Capitalism and Colonialism, Technology and Culture

David Hesmondhalgh

1 MUSIC CULTURE TRANSFORMED?

Starting around 2010 and picking up pace from around 2015, music streaming has spread across the world. The experience of music has been transformed for hundreds of millions of people, and so, too, have the ways in which music is produced and distributed. The term "music streaming" technically refers to providing users access to vast catalogs of content stored on distant servers rather than on their personal devices. One consequence of streaming is that music is now much less embedded in material "sound carrier" artifacts (CDs, cassettes, vinyl records) that are purchased and owned by consumers than it once was.[1] But there's more to music streaming than this, as it mainly operates through music streaming *platforms* (MSPs). Strictly speaking, these are *audio* platforms because they include nonmusical material, including podcasts and audiobooks; but the vast majority of their content is music.

Digital platforms have transformed many fields of social and cultural life, among them health, education, transport, how we communicate, and even how we spend our time. A term now widely used in academic research for the intervention of platforms into various domains, including cultural production and consumption, is "platformization."[2] All digital platforms depend, to varying degrees, on "datafication"—the collection and processing of vast amounts of data. In the case of MSPs, this includes data about where users are located, what they play and compile into playlists, whether they skip tracks, and also about the music (its pace, textures, moods, and much else).[3] Digital platforms also make much use of automated or "algorithmic" recommendation, a key element in their efforts to keep users engaged and discourage subscription cancellation and/or increase advertising revenue.[4] The platformization of music means that its production and consumption are now shaped, and some would even say controlled, by the information technology sector, displacing the consumer electronics industries that were key in shaping recorded music in the twentieth century.[5] Billions of people access sounds via MSPs, many of which are "western" in origin, such as Spotify, Apple Music, and Amazon Music. But some aren't, including Gaana (India), QQ Music (China), and Boomplay (much of Africa, though Chinese-owned).[6] Also vitally important to the circulation and consumption of music are video and social media platforms such as YouTube, TikTok/Douyin, Instagram, and Kuaishou. Music is now shaped by an ecology of platforms that includes MSPs but goes beyond them. Access depends on devices that barely existed thirty years ago: powerful smartphones and laptops, voice-activated speakers, Bluetooth headphones, and so on.

Just as music was the first significant cultural form to be subject to the multiple processes that came to be known as "digitalization," it was also the first major cultural form to be platformized. As so often when new technologies become associated with cultural transformation, music streaming has been controversial, as fans, musicians, journalists, and academic researchers have sought to make sense of sometimes bewildering changes.[7] Music streaming's dependence on information technology—particularly the collection and processing of data—makes it subject to many of the concerns that surround tech power in other domains, such as social media: privacy, surveillance, and compulsive, unrewarding patterns of use.[8] Streaming services are held responsible for making it more difficult than ever for musicians to make a living from recorded music.[9] "Algorithms" are blamed for reinforcing the power of superstars and the corporate entities that work with them, as well as for constraining the musical tastes of consumers.[10] Platformization has more generally been linked, in public debate and academic research, to a deterioration in musical experience: distracted listening habits, the cultivation of an individualized musical experience at the expense of encounters with social difference through music, and the idea that functional forms of musical experience (using music to work out, relax, get to sleep, etc.) are taking over from more aesthetically oriented ones.[11] There have even been claims that music itself has been affected—that it has become bland and unchallenging, that tracks are becoming shorter, limiting musical development, and that musicians and the businesses that fund them are using mere gimmicks to grab the attention of audiences and retain it beyond the thirty seconds needed for a stream to register.[12]

Chapters in this book touch on such developments and controversies in different places around the world. But the main focus of this volume (and this chapter) is to understand the effects of streaming on the circulation of music across nations and continents, and what those effects tell us about power, justice, and inequality in music culture. For there have also been concerns about how platformization might affect the international circulation of music and whether music might increasingly be mediated via "neocolonial" digital technologies originating in the west. The term "music culture" is deliberate. Although "the music itself" (performances, recorded and otherwise) is important to all the contributors to this book, we are also interested in the production, distribution, and consumption of music, as well as evolving relationships of music to society. To put it more concretely, the book does not seek to adjudicate in the often rather simplistic claims about musical decline referred to above. Nor does it provide a series of histories of which genres and performers are most popular, innovative, or controversial in particular countries, and how streaming relates to such trends—though genres and performers do feature throughout. Rather, each chapter addresses instances of change/continuity in music culture in the era of streaming, providing a variety of perspectives from across the world. This reflects the fact that many of the contributors work at the juncture of music studies, cultural sociology, media and cultural studies, and internet studies.

How and in what ways might the dominance of music streaming be (re)shaping music production, distribution, and consumption internationally? Does the global spread of music streaming favor western technologies, business practices, and cultural forms at the expense of those associated with the Global South or the "semiperiphery" of the Global North? What might music streaming, as an instance of the "platformization" of culture, tell us about the relationships between political-economic power and inequality on the one hand, and cultural change on the other? Might music streaming even be a form of "cultural imperialism" or, in more recent coinages, "digital colonialism" or "platform imperialism"? Is it just downright colonialism?

To frame the book's contributions to understanding these and other related questions, I provide in the following section an international overview of the uneven development of music streaming. Crucially, this is embedded in a long-term analysis of the uneven development of music production and consumption since the global industrialization of cultural production and consumption began in earnest in the early twentieth century. Section 3 then recounts the role digitalization and platformization have played in the reshaping (and continuity) of music culture in the twenty-first century. Contextualizing cultural platformization in the *longue durée* is vital because it cannot be adequately understood without reference to the industrialization and commodification of culture, as well as the entanglements of those processes with capitalism, colonialism, and modernity. So, in section 4, I address ways in which historically activists, critics, and researchers have sought to understand international inequality in culture, and in music specifically. I recount how twentieth-century developments were challenged by critics in the Global South and their western allies using frameworks of media and cultural imperialism, and how, in turn, these critical frameworks were challenged by "globalization" theorists. I argue that neither media and cultural imperialism nor globalization theory were adequate ways of understanding international flows. Nevertheless, section 5 uses a typology derived from cultural imperialism theory to consider the degree to which developments in the age of streaming might be considered to be entrenching or qualifying dynamics of neocolonialism, centered on three issues: ownership of the means of production and circulation; flows of cultural products from the west to the Global South, and counterflows in the other direction; and the spread of western practices and habits. I argue that it is the third area—and, to some extent, the first—where neocolonialism is most apparent. As I proceed, I offer brief summaries of the contributions to this volume.[13]

The volume therefore responds to calls for greater attention in studies of popular culture, cultural industries, and music to developments beyond the west and seeks to build on previous efforts to take a more international approach. In media studies, music studies, and cultural sociology, such initiatives—and related ones to move beyond understandings based on "western" or European and North American assumptions and cases—often draw on concepts such as "internationalization," "de-westernization," or "decolonization" to frame such calls. Researchers using these concepts often rightly point to inequalities in academic resources and prestige, as well as to the limited use by western scholars of perspectives and sources from the Global South.[14]

The present book cannot claim to make anything more than a small move toward addressing and correcting such problems. But it does point to international power inequalities and various ways they might be contested and complicated, based on a large number of cases beyond Western Europe and the United States, including numerous authors with strong connections to Asia, Latin America, and Africa, as well as some whose work focuses on Europe and the Anglophone world. In doing so, it also builds on past endeavors to pluralize music studies by considering a wide range of international contexts and understanding them from a reflexive and politicized standpoint.[15] No single volume, even one of this size and scope, can be comprehensive in global coverage. Nevertheless, this book provides coverage of developments in South and East Asia (China, India, Japan), Latin America (Mexico), North and East Africa (Egypt and Kenya), Southern Europe (Italy), and Eastern Europe (Hungary), as well as a study of connected developments in the US core regarding the idea of the "metaverse." It considers the distinctive ways in which music culture might be evolving in the particular regions and nations under analysis, as well as the entanglements of change and continuity apparent there. The last word is given to Yiu Fai Chow, a distinguished cultural studies scholar and Cantopop lyricist based in Hong Kong.

2   THE GLOBAL SPREAD OF THE MUSIC INDUSTRIES

European colonization of land and people involved musical colonization, too, in particular the spread of "church music, choral anthems and light music for dancing and entertainment."[16] As the musicologist Kofi Agawu has shown, these musics, especially hymns, helped spread the European musical system of tonality in Africa, imposing unfamiliar rhythms and textures and disregarding indigenous musicality. Such music "defined civic and religious life in communities" across Africa. Agawu recognizes that tonal music in some ways "enhanced" civic and religious life. Yet he is also clear that it constituted "musical violence of a very high order, a violence whose psychic and psychological impacts remain to be properly explored."[17] Cultural historian Michael Denning expands on how colonial musical practice "instituted new disciplines of the body—new ways of singing, of dancing, of marching, of playing instruments" and also reshaped how bodies were articulated with each other in the formation of marching bands and church choirs.[18]

The twentieth-century international recording industry developed out of Euro-American capitalism and colonialism. Consumer electronics businesses based in Western Europe and the United States developed, manufactured, and marketed "hardware" devices intended for business and domestic leisure use; they financed, produced, and marketed recordings to be played on those devices.[19] Two main companies emerged: Victor in the United States and the Gramophone Company in Europe. Seeing the benefits to be gained from internationalization, these western businesses divided the world between them in a cartel agreement of 1901: Victor got East Asia and South America, the Gramophone Company got the rest of Asia, Africa, and Australasia. Representatives of the latter company toured the planet, making recordings of local musicians; branches and agents were established across the world.[20]

The spread of recording and playback technologies, along with business processes, from the European and North American core to global peripheries, was dependent on intellectual property frameworks of patent and copyright developed in that core, and subject to a series of international agreements signed in an international capitalist order centered on Euro-American colonialism. The crucial foundation for the global spread of copyright, the Berne Convention of 1886, was developed by colonial powers, who could, by default, incorporate their colonies and territories. In one of many later revisions and extensions, Berne was applied to sound recordings in 1908.[21] This and other expansions of the international copyright regime meant that, by the mid-twentieth century, copyright was generating significant trade gains for cultural exporters. In the words of two critical analysts of intellectual property, by the time many countries shed their colonial status, they were confronted by a Berne system "that was run by an Old World club of former or diminished colonial powers to suit their economic interests," with legal and regulatory expertise located almost entirely in the colonizing states.[22] Even the very stuff of recordings was a colonial product: from around 1900 until the introduction of vinyl in 1950, discs were mainly made from the raw material of shellac resin secreted by the lac insect, harvested and processed in India under terrible, exploitative conditions.[23] Another victory for western rights-holders was the international spread of performing or "neighboring" rights, initially enshrined in the Rome Convention of 1961.[24]

Industrialization and commodification had their limits. Vast swathes of musical activity continued to be conducted in "informal" economies and low-cost settings (local live music venues, dance events, weddings, etc.). This was the case in the Global South and away from the gaze of established music-industry organizations in wealthier countries too. To overstate the degree of metropolitan power exerted by colonialist capitalism in this system would be to efface developments beyond the west. Beneath the top layer of international Anglophone success, there were also regional power centers, such as Egypt's domination of the Arab recording industries from the 1930s onward, and Taiwan and Hong Kong's key role in Chinese-language music from the 1970s.[25]

A further complexity is that it would be very wrong to understand the globalization of music in the twentieth century as an unqualified and unresisted imposition of the music of the west on the rest of the world. In most places, local repertoire drawing on local styles, in local languages, competed with Anglo-American international repertoire for local attention and success, in some places very effectively.[26] Some of these successful local styles were exported successfully to the west, and globally, hence the international popularity of Argentinian tango, Jamaican reggae, Brazilian samba, and many other styles. These musics often had their origins in racialized, diasporic, and marginalized communities. The same was true of much of the most commercially successful Anglophone popular music, often based on African American styles (jazz, R & B, soul, funk, hip hop) and/or on white appropriations of those styles.[27]

Michael Denning has shown how the international spread of the gramophone in the 1920s "amplified a musical revolution that was already taking place in the urban streets and music halls around the world," based on infrastructures of transport and communication (steamships, railways, telegraphs), "moving commodities and people across and between empires."[28] For Denning, these musics "not only captured the timbres of decolonization":

> The emergence of these musics—hula, rumba, beguine, tango, jazz, samba, marabi, kroncong, taraab, chaabi—*was* decolonization. It was not simply a cultural activity that contributed to the political struggle (though there are cases of musicians taking political stances and actions); it was somatic decolonization of the ear and the dancing body. . . . The global soundscape was decolonized by the guerrilla insurgency of these new musics before the global statescape was reshaped.[29]

Another factor complicating notions of empire as a force of globalizing homogenization was that jazz and rock may not have been quite so hegemonic in their global spread as we suppose. After all, significant barriers were put into place by states, whether "Communist" or religious, to prevent or discourage the import of "western" sounds, though this may have added further to the prestige of rock among younger and highly educated audiences in the west and the Global South.[30] Some governments put considerable efforts into supporting local production against the might of the Anglophone music industries, including in some non-Anglophone countries and regions, such as France and Quebec, the imposition of radio "quotas" requiring broadcasters to play a certain percentage of local or local-language content. But in any case, older, local forms—even if some were, in reality, hybrids incorporating elements of earlier western colonial exports—often remained resilient, regardless of the existence or absence of such barriers, often thriving in informal economies as well as via more formalized routes.[31]

Moreover, western technologies continued to have unforeseen consequences. One of the most important cultural technologies exported from the west around the world, the audiocassette, which was introduced in the 1960s, eventually served as the basis for the development of domestic music industries in the Global South, which began to challenge the dominance of western record companies. Writing in 1999, music industry historians Pekka Gronow and Ilpo Saunio remarked, using the terminology of the time, that "the multinational companies have still not gained the foothold they had in the Third World before the advent of the cassette."[32]

Nevertheless, as global trade in culture grew rapidly in the late twentieth century, ever more powerful and multinational corporations, based on "synergies" (multiplying interactions) of hardware and software production, dominated the music industries and acted as formidable lobbyists for extension of copyright duration and for more rigorous enforcement. In the "globalizing" boom of the 1980s and 1990s, revenues from the music industry's fiendishly complex system of "rights" rocketed. Meanwhile, international governance of intellectual property shifted to bilateral agreements and to the newly powerful World Trade Organization.[33] Boosted further by sales of compact discs and the expansion of income from neighboring rights, the recording industry achieved a new peak of profitability and power in the 1990s, especially in Western Europe, North America, Australasia, and Japan. Even so, the most popular global acts were mainly Anglo-American artists, whose works were exported from the metropolitan hubs of Los Angeles, New York, and London, and the most popular internationally circulating genres were identifiably western ones. Bob Marley, who died in 1981, remains the only genuine global superstar to emerge from the Global South.[34] English was extremely dominant as the premier language of internationally circulating pop. Artists seeking international success whose first language was not English were expected to write, record, and even publicize their music in English anyway.

3   THE DIGITAL AND STREAMING ERAS: A NEW SYSTEM EMERGES

Even as the western-dominated music industries achieved unprecedented wealth and international reach in the 1990s, threats were already visible on the horizon. An extraordinary proliferation of information technology companies had evolved out of the United States' vast systems of defense-based R & D, university research and finance capital. As just one of its many insertions of computerization into various domains of life, but enabled by music's low bandwidth, this rapidly expanding IT sector had developed and marketed technologies allowing for easy copying, uploading, and sharing of music.[35] In the process, it undermined sales of recorded music objects (vinyl discs, cassettes, CDs) that had been the basis of the music industries; rights-holders framed such "unauthorized" copying and sharing as "piracy." Across the industrialized world, revenues of music recording and publishing companies plummeted during the 2000s. For the western music industries, capital investment, the lifeblood of business corporations in the age of finance capitalism, collapsed. Elsewhere, nascent recording industries, in many places already subject to problems deriving from the chaotic aftermath of colonialism, including "commercial piracy," were further stymied.

Under pressure from cultural industry lobbying and governments, the IT sector offered a temporary bandage for the music industries' wounds in the west via new hardware devices and software systems that encouraged consumers to pay for digital downloads by making copying and sharing much more difficult, most famously Apple's iPod hardware (2001) and iTunes music store software (2003). At around the same time, however, a more robust solution to western music industry problems was emerging in the form of a bigger development: the emergence, across many aspects of economic, social, and cultural life, of the digital platform.

The openness of early internet and web architecture had been celebrated by digital optimists for its ability to generate multiple possibilities of use. However, this very openness made peer-to-peer systems vulnerable to new layers of protocol being imposed on top of them.[36] Developments in the collection and processing of data, perhaps most notably at Google, led to new business models that matched consumers to advertising niches.[37] Some consumers were always going to value convenience and security above generativity and openness, and the former values were heavily promoted by tech companies purporting to offer them via the emerging form of the digital platform. Legal and regulatory shifts sought to extend intellectual property into the proliferating new realm of "information."[38] While "platform" had been used to refer to combinations of hardware and operating systems (such as "the PC") and to websites (such as eBay) since the 1990s, the digital platform emerged from later developments. By the mid-2000s, a number of innovations in the information and communication technology (ICT) sectors (i.e., telecommunications, and computer hardware and software) provided the basis for the launch and growth of digital platforms: state-supported rollout of fast broadband, the launch of "app stores" that enabled easy access and subscription, and the growth of mobile telephony networks, often offering such services for free.[39]

Search engines and social media (then called social networks) were in the vanguard of platformization and datafication, but in music, start-ups such as Spotify and Deezer led the way, building on earlier ventures such as Rhapsody (launched in 2001) that had borrowed the underlying technologies and interface designs of "pirate" sites enabled by the generativity of internet and web architecture.[40] These later start-ups, emerging in the 2006–8 period, were able to build closed "trusted system" architectures on top of internet infrastructure, making it impossible for all but the most sophisticated users to tinker with them, by contrast with earlier phases of the web.[41] In order to avoid the legal challenges that had closed down unlicensed sites such as the famous Napster, these operators, by now labeled "streaming services," followed the example of Apple's iTunes by working closely with the major rights-holders to gain licensing agreements that the record companies and publishing companies had previously refused to grant. No streaming service could compete without gaining access to the most popular existing content. As agreements were signed, all the main platforms came to be centered on the same core repertoire, the majority of it licensed from what by now were three major multinationals (Sony, Universal, and Warner), along with larger independent record companies.[42]

Adoption of platforms gathered pace, and revenues from both advertising and subscriptions grew. Spying opportunities to enhance their "offers," three tech giants (Apple, Amazon, and Google), flush with excess cash, entered the western streaming service market between 2013 and 2016, alongside Spotify, Deezer, and some smaller players.[43] By now, the terms "music streaming platform" and "audio platform" were being widely used to describe the streaming services, replacing earlier terms such as "music in the cloud" and existing alongside industry nomenclature such as "digital service provider" (DSP). These MSPs were increasingly accessed via mobile phone and laptop apps rather than websites, offered as part of smartphone packages or the bundled services of the IT giants, often on free trials. Subscription prices were low: in the west, vast amounts of musical content were available for less than ten dollars or ten euros per month, alongside free tiers for those willing to tolerate advertising. In the Global South majority world, services were operated as part of relatively low-priced mobile telephony packages, and sales of ringtones often operated as a key moment of transition to digital music, as Andrew Eisenberg recounts in his chapter for this book, where he also explains how a mobile-phone money transfer package helped lay the basis for the platformization of music in Kenya.

The formidable licensed repertoire of the multinational corporations and the larger independents was increasingly supplemented on MSPs by content that could be easily and cheaply uploaded by smaller businesses and even by "self-releasing" or "DIY" (do-it-yourself) musicians.[44] New digital intermediaries enabled uploading and tracking of data and, in some cases, also offered marketing and other services.[45] Because licensing is usually carried out on a transnational basis, tracks uploaded in one country will often be available in most or even all other countries and territories with access to the streaming service concerned (by contrast with the more complex systems of licensing on video platforms). This contributes further to the politics of abundance surrounding MSPs, and it may also have contributed to a reconfiguration of international musical flows, involving a greater transfer of sounds from the Global South to the minority world core of the north (see below).

In 2015, Spotify reported that thirty million audio tracks were available on its service; by 2024, the figure was well over one hundred million, the great majority of which was music. Data company Luminate detected 184 million audio tracks on streaming services in 2023, with approximately 120,000 tracks being added every day, much of it now produced by generative AI technologies (a potential source of crisis). This is clearly an unprecedented abundance—though most tracks are rarely played, and indeed a vast number are not played at all.[46] In the wake of such abundance, debate has continued about the extent to which recorded music in the age of streaming remains a "winner take all" superstar economy, and the degree to which streaming makes it possible for artists down the "long tail" to thrive. There does seem to have been a small move down the long tail, but this seems to be the result of a conscious reaction by some MSPs, or at least Spotify, to expressions of public concern; this makes clear that MSPs have some power to adjust distribution via their recommendation systems. More musicians than ever before earn revenue from recorded music, but it is harder than ever to rise above the fray and become successful enough to earn a sustainable living.[47]

We will see in this collection a number of international takes on how streaming might be reconfiguring the continuing struggle for musicians and others who work in music (managers, recording and publishing company staff, tour managers, PR people) to earn a living from music. Onur Sesigür provides an insider account of how musicians in Istanbul, the center of Turkey's expansive music industry, often turn to creating music for advertising as a way of earning their living in a city marked by expensive rents, relating this to long-standing debates in music culture about "selling out"—that is, modifying or abandoning aesthetic commitments (often with underlying ethical or political elements) for financial gain.[48] Zhongwei (Mabu) Li and D. Bondy Valdovinos Kaye consider the importance of digital self-releasing in China, outlining a shift from a situation in the early 2000s—when self-releasing allowed musicians to evade the strictures of the established industry, in a period of open internet infrastructure and loose copyright oversight—to the circumstances that prevail now, in the era of platformization, where agency and artistic expression are significantly constrained, in a system of datafied music distribution, and institutionalized copyright. Providing an overview of the platformization of music in India, Aditya Lal points to the rise of a recorded music industry based on regional and nonfilm music, displacing to some extent the Bollywood film industry (see Lal's chapter for definitions) that was previously extremely dominant. However, echoing a common theme in studies of cultural labor in the digital era,[49] Lal points to how the seeming multiplication of options seems to have created false hope among independent musicians—a category that was practically nonexistent in the predigitalization era—who persistently risk investing their own finances in music production, publishing, and marketing in hopes of attracting the attention of listeners and the powers that be. Finally, Emília Barna shows how musicians internalize the expectation to be self-dependent and "entrepreneurial," but reluctantly so. In the case of Hungary, their creative autonomy depends on factors such as background, gender, and political connections. Building an international presence from a semiperipheral base such as Hungary requires enormous effort.

It soon became apparent to MSPs that they had to find ways to allow their users to navigate their way through the abundance of music they were making available, and in a way that served the ideology of personalization that had developed since the emergence of personal computing in the 1990s. Three main techniques emerged. The first was borrowed from sharing sites and digital download stores such as Apple's iTunes: the use of "playlists" as a way of organizing musical content, so that genres, albums, eras, popularity charts, and the work of artists were often or even usually presented in that form. In turn playlists became key mechanisms by which artists and songs become known to audiences, leading to battles over inclusion. There have been various helpful studies of playlists and the industry processes behind them, but Francesco D'Amato analyzes the form these processes take in the contemporary Italian industry, with particular attention to "playlist pitching"—the efforts by intermediaries to get music on to playlists. In a country where local repertoire is dominant, it is striking that decisions about playlisting are made by a tiny number of employees of international MSPs: two curators and two label relations staff at Spotify (which takes at least 70 percent of the market) and just two at Apple. D'Amato provides the most detailed analysis available of the interactions of local platform employees with local music industry professionals from the majors, showing that playlist placing for emerging artists can be highly contingent, subject to chance about whether established artists are releasing anything at around the same time.

Emília Barna provides another illuminating account of the mediation of relations between MSPs, (small) labels, and musicians. She shows how in a "semiperipheral" market such as Hungary, musicians' income and opportunities are highly dependent on two factors: first, platforms' geographical policies, in particular local representation in the form of offices or local playlists; and second, the role of distributors in the global market and the deals between platforms and collecting societies. But Barna also shows that local players, in particular nonprofit collective organizations serving musicians' interests, can work for better working conditions by protesting existing platform policies.

The second technique for managing abundance was a mix of such humanly "curated" recommendation with automated versions, as discussed above (see note 4). Raquel Campos Valverde shows how automated recommendation is dependent on opaque taxonomic systems that potentially encode western "biases" into international music culture. She outlines how these taxonomies work, arguing that metadata coding standards currently followed by the music industry rely on inadequate understandings of genre classification or sound analysis, particularly regarding non-western musics and much western music beyond its central canons (popular and classical).

The third technique is more obvious: the interfaces of MSPs play a key role in shaping musical experience. These interfaces have converged on a set of strikingly homogenized design conventions built on playlists and recommendations, with a set of recurring themes across pretty much all western platforms, though the emphasis on specific elements varies by user. Alongside genres, artists, eras, and albums, playlists based on "function" (working out, getting to sleep, waking up) and mood ("good energy," "sad songs," "peaceful piano") achieved a new prominence compared with the era when retail, radio, and television were the key means of presenting music. This, in turn, generated new controversies about musical functionalism and the commodification of mood.[50]

The result, in wealthier parts of the world, is a new system centered on convenience, abundance, and relative cheapness for consumers, involving largely automated personalized recommendations and the collection and processing of huge amounts of data. As with the rest of the digital world, finance capital is at the heart of the system, and vast tech corporations exist alongside myriad start-ups seeking riches through innovations. In this new system, music is often (though not always) experienced in highly individualized ways: in particular, it comes to many people via headphones connected to mobile phones and laptops. Musical experience is also closely integrated with other aspects of digital media, such as social media and short video platforms, and shares the push toward personalization apparent there—with implications for notions of musical community that are as yet underexplored and poorly understood. As Jeremy Wade Morris shows, streaming is also connected with games and emerging immersive technologies such as the "metaverse." For MSPs, the metaverse is an opportunity to diversify by appealing to new, younger users, potentially extending the concept of streaming toward a more diffuse musical experience in a virtual space.

There is great variability in the prevalence of streaming across different countries. The wildly unequal wealth of nations is a major factor, partly because it influences the extent of digital infrastructure, broadband connections, credit card access, and affordability of mobile data. But it is not the only factor, and perhaps not even the main one. It is notable that some of the wealthier countries have adopted streaming much more slowly than others. Germany is one example. Another is Japan, a "laggard in the adoption of internet-based music, especially streaming," as Noriko Manabe puts it. Manabe explains that Japan was slow because the music industry successfully maintained the popularity of CDs and resisted the onset of platformization. This delay may have been related to the lesser power of the information technology sector in Japan, where consumer electronics have been so central to the economy. Such factors suggest that the pace of streaming adoption is influenced by country-specific conditions, including the specific makeup and lobbying power of the traditional music industries. Manabe shows that even though it was adopted "late," relative to Japan's wealth and digital connectedness, streaming has begun to transform music culture in the country by breaking the dominance of TV-generated idol pop and leading to the rise of niche or unusual acts, including acts based on musical avatars.

Conversely, some less wealthy countries have seen significant adoption of streaming. Rodrigo Gómez, Ignacio Gallego, and Argelia Muñoz-Larroa analyze the case of Mexico, which generated a remarkable 94 percent of its music revenue from streaming in 2023. In that year, 57 percent of listeners recently "engaged with" paid subscriptions, higher than in Germany and the United States.[51] A key factor here, as well as in other Latin American countries such as Brazil, where streaming has seen high uptake, is the widespread use of smartphones and a culture of spending extensive time on them. Of course, global inequality means that, in many countries, streaming is confined to a minority. The world's twelfth-most populous country, Ethiopia, is one such example; it also has no functioning copyright collection society at the time of writing. But Andrew J. Eisenberg shows that a different situation prevails in neighboring Kenya, where a professionalizing industry increasingly based on streaming has come to displace the "piracy" of earlier generations.

Robert Prey and Seonok Lee have identified other important variables in music streaming internationally, besides the degree to which streaming has been adopted in different countries.[52] One is the extent to which platforms are integrated with the music industries. In some countries, the level of integration is low, as the platforms, owned and operated by tech companies, are separate from the music companies that license the most popular content to them. This is the case in Europe and North America, where take-up of streaming is high, but also in Nigeria, where users access both global platforms and the Chinese-owned company Boomplay (the biggest streaming service in Africa), despite overall streaming adoption being relatively low. By contrast, South Korea has a very high level of engagement with music streaming, but the leading platforms there (such as Melon) are also involved in the production of music.[53] Similar dynamics prevail in China, as Shuwen Qu and D. Bondy Valdovinos Kaye show. However, Qu and Kaye also reveal further levels of specificity in analyzing the phenomenon of hit tracks on the MSP Douyin, widely used in China. They show that these hits represent a shift in the Chinese music business, one that appears to differ from star-making practices across much of the world, emphasizing instead the tracks themselves and downplaying the importance of performer identity. While there are precedents for such practices in all recording industries, the shift toward anonymity, driven by the rise of short video, once again shows the intellectual bankruptcy of assuming any kind of homogenous international model for music production and consumption in the age of streaming—and probably in any era. Another element of variability identified by Prey and Lee is the degree to which streaming platforms and music business companies are owned by local versus overseas businesses—an issue to which I return in section 5.

4   FROM CULTURAL IMPERIALISM TO DIGITAL COLONIALISM?

How might we read this combination of platformization and variability? How to interpret the effects of streaming on musical production and consumption through the lens of global inequalities? To address these matters, let me now return to the task of putting digitalization and platformization in the context of the long-term industrialization and commodification of music, as well as the basis of these various developments in capitalism and (neo)colonialism. I shift from discussion of more specific, concrete cases to consideration of how the internationalization of music on capitalist-colonialist grounds was promoted and critiqued. I want to pay particular attention to the concepts used by critics, in particular their use of ideas of imperialism and colonialism, and their relationship to culture and music.

The growth of the music industries internationally in the twentieth century was only one manifestation of the industrialization and commodification of culture; there were parallel developments in other sectors, notably news, film, and television.[54] The immense cultural-economic power of the United States was a major factor. The dissemination of media, along western lines, became linked to processes of "modernization" and "development."[55] From the '40s to the '90s the spread of US cultural goods was intimately linked to the Cold War goal of persuading overseas listeners of the superiority of "American" ways of life.

These international cultural-economic inequalities sparked a reaction from anticolonial movements, postcolonial states, and their progressive allies in the western core. They were often understood via a cluster of concepts such as "cultural imperialism," "media imperialism," and "cultural dependency." These terms were rather loosely conceptualized in academic research, often acting, in Annabelle Sreberny's words, as "evocative metaphors" rather than as a basis for sustained efforts to understand relations between culture, media, imperialism, and colonialism.[56] Nevertheless, activists and intellectuals guided by these ideas drew attention to important dynamics: flows of media and cultural products from the west to the non-west; ownership of the means of cultural production by western-based businesses; the inculcation of modern and often metropolitan practices and habits that originated in the west; and threats to Indigenous and "traditional" modes of living—a set of issues to which I return below.[57] At the United Nations, there were conflicts over claims of the need for a New World Information and Communication Order that allowed space for "the Third World" (leading to the eventual withdrawal of the United States and the United Kingdom from UNESCO in 1984–85). Particular concern was directed toward flows of news and their potential reinforcement of US military power, the long-standing domination of Hollywood, and the spread of consumerism via advertising. Remarkably little of this work paid any attention to intellectual property, despite the centrality of copyright to the international industrialization of culture on western, often colonial or neocolonial terms. Very little of the work addressing media and cultural imperialism made any reference to music, but cultural activists, ethnomusicologists, and others expressed parallel concerns about the threat posed by industrialization and modernization to traditional and folk musics.[58] A later wave of critique from activists and some researchers was directed at the phenomenon of "world music," with particular attention to the appropriation of non-western styles by superstar musicians (the most famous case being Paul Simon's album *Graceland*, recorded in South Africa) and the labeling of any music beyond the Anglo-American global core in simplistic and sometimes ethnocentric terms.[59]

From the early 1980s to the late 1990s, however, academic opinion began to turn against the media and cultural imperialism thesis, often on the grounds that it overstated or simplified relations between west and non-west. Critics began to point out the rising economic and cultural power of "newly industrialized countries," such as South Korea, and the cultural power of some former colonies—for example, the significant global presence of India's film industry.[60] There was an increasing sense that critiques of media and cultural imperialism that depended on notions of cultures "uncontaminated" by western influence were problematically nationalist or even naive. Instead, cultures came to be seen as fundamentally hybrid in nature and often desirably so.[61] Many commentators pointed to the way that exports from the west to the east, north to south, core to periphery, were subject to appropriation and creative mixing on the part of the populations that received them. Having been rather absent from considerations of media and cultural imperialism, popular music featured fairly prominently in some critiques of the assumptions behind the cultural imperialism "thesis." When I wrote above of the importance of not slipping into a portrayal of popular music's international flows as simply an imposition of the west on the rest, I was drawing on some of the most thoughtful of these critiques.[62]

As digitalization took shape with the emergence of the internet and web in the 1990s, however, concerns of activists and governments began to shift towards what was starting to seem like a more urgent set of priorities. Digitalization had sparked new hopes for modernization and development from the late 1980s onward, but activists quickly drew attention to the enormous "digital divide" between and within nations, and their potential effects in domains such as health and education. Two World Summits on the Information Society in 2002 and 2005 pitted technocratic discourses of government against civil society groups stressing the serving of human needs.[63]

When the 2000s and 2010s saw the evolution and spread of digitalization, researchers focused on media and social media began to adopt concepts such as "digital colonialism" and "platform imperialism" to characterize the activities of US Big Tech corporations, which operated in alignment with the US state, and to highlight damaging consequences in the Global South.[64] The critical thrust of this work was essential, and the most distinguished contributions represented important advances.[65] As in the era when notions of media and cultural imperialism prevailed, however, there sometimes appeared to be a slippage toward using these terms as "evocative metaphors" rather than as the basis of a thorough conceptualization of colonial and capitalist power. Moreover, given our concerns, the lack of engagement with *culture* in critical scholarship that draws on terms such as digital colonialism and platform imperialism is striking. By "culture" in this context, I simply mean the domain of art, entertainment, and the self-expression of individuals and collectivities. This sidelining of culture is apparent in the lack of attention to older concepts such as media imperialism and cultural dependency. Instead, the focus in such critical research on international digital inequality has been overwhelmingly on concepts of information and communication in more general terms, often approached via analysis of social media, and emphasizing sociotechnical concepts such as data and algorithms rather than culture in the above sense. Unsurprisingly, then, music is mainly absent from such treatments, as it mostly was in the first wave of media and cultural imperialism writing and activism.[66]

Another striking absence in research using concepts such as digital colonization and platform imperialism is the role of intellectual property.[67] Critical legal studies of various kinds in the 1990s and early 2000s had offered important perspectives on how global IP law served the interests of the west and brought about active harms in the Global South.[68] Such studies rarely used imperialism and colonialism explicitly as frames, but they were certainly concerned with international justice and inequality.[69] Egregious developments such as the effects of patent control on health outcomes were understandably prevailing objects of attention, and there was also an important strand of research investigating injustices deriving from the spread of copyright.[70] Yet copyright seems so far to have evaded careful scrutiny in work on digital colonialism and platform imperialism.

The above comments are not intended to belittle activism or research under the banners of digital colonialism, platform imperialism, and other related concepts, especially not the inspiring initiatives aimed at countering harms inflicted on the Global South by monopolistic tech corporations such as Facebook.[71] Activists need metaphors to label and promote their work. But the neglected task of understanding relations between culture, capitalism, and colonialism might benefit from greater analytical precision than concepts such as "digital colonialism" have so far been able to provide. This partly reflects the low status of culture in public policy and social justice activism, perhaps understandable given overwhelming imperatives regarding poverty relief, health, and education. But culture matters, too, in different ways—and so does music.[72]

5   MUSIC AND INTERNATIONAL INEQUALITY IN THE STREAMING ERA

A challenge then for research on the relationships between music, capitalism, and colonialism, when digitalization and platformization are brought into the frame, is how to comprehend those relationships in a way that recognizes the specificity of music as a domain of human life (sidelined by nearly all cultural imperialism and digital colonialism theorists), while addressing the technological, legal, and political-economic forces shaping it (neglected in music studies).

One way to approach those relationships might be to temporarily stand back from older concepts of cultural imperialism and globalization, as well as newer ones such as digital colonialism and platform imperialism, and instead simply to ask the following Very Big Question: In what ways is music in the present conjuncture bound up with contemporary capitalism and colonialism? But that is an entire research program, rather than something that can be answered in a single book—let alone a book introduction. Here, in keeping with the particular gathering of expertise represented in this volume, I return to the taxonomy of issues drawn from cultural imperialism theory of the late twentieth century to assess more recent developments in the age of streaming and how they might be in the process of being reconfigured. (However, as I have tried to make clear, this is not to endorse entirely the assumptions of the late twentieth-century cultural imperialism "thesis.") First, to what extent is the ownership of the means of cultural production by western-based businesses still apparent? Second, how might this relate to flows of media and cultural products from the west to the non-west, and counterflows from Global South to the minority world? Third, to what extent is music (and music streaming) bound up with modern and often metropolitan practices and habits that originated in the west? How are these issues reconfigured in the age of streaming?

Beginning with ownership, as discussed above, the contemporary music industries are centered on two parallel oligopolies: the tech companies that own and operate the means of circulation, and the recording and publishing industries that control the supply of most music. The tech side of this equation can be conceptualized as a set of layers, with infrastructure at the base, hardware in the middle, and the consumer-facing MSPs at the "top" end. Three of the five giants constituting the famous GAFAM tech oligopoly—Google, Apple, and Amazon— have enormous international presence, while Spotify, the largest standalone operator, is the biggest of all, despite being reliant on massive, mainly western corporate finance. Hardware, such as chip manufacture, is somewhat internationally distributed. As for infrastructure, as Dwayne Winseck has shown, the extent to which western tech companies dominate this layer is sometimes overstated: Telecommunication companies are just as important as tech companies in terms of information infrastructure, and many of them operate out of the Global South, including India and China.[73] Meanwhile, the content provision side of the industries are dominated by three corporations, one Japanese (Sony), one US-based (Warner), and the biggest of all, Universal, which has its corporate HQ in the Netherlands and its operational HQ in California. These companies still account for more than 70 percent of global music revenues, and there are also large independent companies operating out of the west that account for quite a bit of the rest. So, although we need to note some qualifications, the ownership picture is still extremely western-dominated.

In some countries, the dominant force in bringing MSPs to millions of people has been western companies with highly international reach: Spotify was available in 184 countries and territories as of 2024, while Apple Music was available in 167 in 2020.[74] Overall, at the time of writing, Spotify accounts for around a third of global music streaming revenue. In some places, local streaming companies compete with the western tech companies, such as the Beirut-based streaming service Anghami in the Middle East and North Africa region, as discussed in Darci Sprengel's chapter. She analyzes the serious struggles faced by Global South alternatives such as Anghami to compete with the major western platforms. In its efforts to extend beyond the Arab world, Anghami is considered "too local" by the international music industry and its investors. However, in its need to bring at least some international repertoire to Arab audiences, its business model has become overextended, and the platform faces critique from local musicians for its poor rates of payment.

Often, these local competitors developed out of local telecoms and mobile phone sectors, such as Boomplay in Nigeria, as discussed in Aditya Lal's chapter, and Gaana and JioSaavn in India (see also Eisenberg's reference to his own previous work on this). And in some countries, the western platforms barely exist, most notably in China, where Chinese platforms dominate, in particular three services operated by the Chinese tech giant Tencent. Of the western music platforms, only Apple Music has a presence in the country, based on the popularity of the iPhone among upscale users, but even Apple achieves only 5 percent of the Chinese music streaming market. And while streaming in China is based on the fundamental features of the streaming system discussed above, it takes very different forms in terms of how it interacts with the music industries compared to the west. Zhongwei (Mabu) Li and D. Bondy Valdovinos Kaye unpack some aspects of that specificity, investigating how self-releasing Chinese musicians have interacted with digital platforms and experienced a significant loss of agency in recent years as these services impose new conditions on their users. And in their chapter, as already indicated, Shuwen Qu and D. Bondy Valdovinos Kaye show how short video platforms are altering the Chinese industry by placing a new emphasis on hit tracks, with much less emphasis than before on the identity of performers—a move away from the star systems that have traditionally supported the recording industry in both the west and post-1989 China.

However, on the second issue—flows of products—the picture is altogether more complicated. One measure of such dominance is the proportion of locally consumed content that is locally produced—for example, how much of the most popular content in any non-western and semiperipheral country is produced within that country. Another is the degree to which the products of non-western countries beyond the west achieve an impact in the west—which are often the most prestigious and lucrative global markets.

In some respects, international music flows in the age of streaming continue to show signs of cultural domination associated with colonialism. The United States still accounts for significant portions of the most popular content in many countries, across a number of genres, as shown by one of the most comprehensive studies of music streaming flows, which examined all Spotify data between 2014 and 2019.[75] However, the same study suggested that while the United States accounts for the largest fraction of music in a number of genres, "preferences for local content have increased through the streaming era, and that trend is consistent across different genres, listener age groups, and registration cohorts."[76] There are also signs that the dominance of the English language in global pop is declining as streaming spreads: English language music's share of the top ten thousand on-demand tracks on streaming services globally fell from 67.2 percent in 2021 to 56.4 percent in the first half of 2023.[77] European markets have seen the dominance of English-language products of Anglophone countries diminish: the most popular content in Italy and Poland, for example, is overwhelmingly Italian and Polish.[78] The same is true of many other countries internationally, such as Brazil.[79] Domestic genres, such as gengetone, the style of Kenyan hip hop discussed by one of Eisenberg's informants in his chapter on streaming in that country, have achieved significant popularity within their countries of origin, challenging and even outstripping the popularity of international repertoire. Furthermore, a number of genres have thrived internationally over the last ten years, the era of the dominance of streaming. Among them is the Latin pop discussed in Gómez, Gallego, and Muñoz-Larroa's chapter on Mexico.[80] Korean pop has become a global phenomenon.[81] Significant numbers of African artists are achieving global success for the first time, and the umbrella term "Afrobeats" is used for a range of musical styles with notable R & B, rap, and dancehall influences.[82] Across the world, hip hop has served as the basis of local variants and hybrids that have achieved significant popularity in local markets, serving as a source of immense creativity and self-expression, especially for Afro-diasporic communities, but for many other groups too.[83] Spotify claims that nearly a quarter of all streams on its service globally are of music that it categorizes as hip hop.[84] The role of streaming in the success of the performers and genres involved would need to be assessed on a case-by-case basis. But it is hard to see that streaming reinforced the power of hegemonically white or Anglo-American musical forms.

Might these developments be understood as contemporary equivalents of Denning's "vernacular phonograph musics," a century later? Can we think of Afrobeats and hip hop variants as vernacular streaming musics, fresh examples of new waves of decolonization, as western hegemony comes under increasing scrutiny around the world? Anyone tempted to dismiss the suggestion outright should be aware that supporters of decolonization in the early twentieth century found it difficult to hear the earlier waves of internationally circulating musics as a decolonization of the ear. Denning points to the "profound gap" between decolonization as a political revolution—one involving the attainment of political independence and the takeover of state apparatuses—and the iconoclasm of cultural revolution: "Anticolonial political activities and thinkers were often tone-deaf when hearing these new musics."[85] He also points to how it took many years for the recordings of the 1920s to spin out their effects, "remaking . . . the very structure of feeling, as new sensibilities and new aesthetics became new ways of living."[86] At the time, those musics were objects of suspicion because of their entwinement with colonial musics, their commercial nature, and the sense that they were imitative—even embarrassing—acts of colonial mimicry; in particular, they were understood as variants of jazz.[87] One can hear the same kinds of suspicion in dismissals by contemporary anticolonial intellectuals of genres such as reggaeton and local variants of hip hop—except in those very rare cases where the artists involved incorporate explicit political themes into their lyrics.[88]

Regardless of whether the claim for these newly circulating musics to be understood as forms of cultural decolonization can be sustained, the increasing emphasis on local content, the declining importance of English-language music, and the rise of genres not easily associated with the white Anglo-American imperial center of the music industries surely complicate any effort to see streaming as a digital version of cultural imperialism or colonialism.

6   MUSIC STREAMING PLATFORMS AS THE SUPERMARKETS OF CONTEMPORARY GLOBAL MUSIC CULTURE

I want to suggest that it is the third element of the typology I have borrowed from earlier cultural imperialism research where the capitalist-colonialist nature of streaming is most apparent: that the global spread of streaming, centered on the digital platform, involves the dissemination of a set of relationships to music, and ultimately to everyday life, that are capitalist, western, and ultimately colonialist. An analogy with the global spread of the supermarket might be an illuminating way to explore this claim.[89] Supermarkets and streaming platforms both base their appeal on offering cheap and convenient access to abundance. And just as the core of contemporary food culture for billions of people is the supermarket, platforms are now the economic and cultural core of music in much of the world. Any particular supermarket offers goods and facilities that are very similar to those offered by other supermarkets (aisles, carts, checkouts—even the categories used to describe each aisle recur with only minor variations across different businesses). Similarly, MSPs offer pretty much the same musical content as one another, supplied by the big music companies (exclusive podcasts are only a small part of their content). Rather than aisles, MSPs arrange their audio goods mainly via carousels of geometric shapes, through which users must scroll rather than stroll. Just as supermarkets label their aisles in broadly similar ways, the leading services (Spotify, Apple, Amazon, YouTube, and even lesser rivals such as Tidal, Deezer, and SoundCloud) categorize their carousels using pretty much the same labels as one another. Music is largely organized by artist, genre, album, popularity charts, era, or decade, with some space for new releases. The main novelty is the addition of mood and function categories: chill, work out, relax, energize, get to sleep, et cetera.[90]

The fact that it is hard to imagine other ways of categorizing food and drink in a retail environment is testament to the cultural hegemony of the supermarket as an economic and cultural institution. And MSPs now frame the possibilities of musical experience in a particular set of ways that are also coming to seem natural. The convenience and relative cheapness of supermarkets make them more attractive to most consumers than other, often more expensive and time-consuming ways of accessing food. The same is true of music in the age of platforms. Instead of spending time and money getting to a record shop, users can access millions of tracks with a few interactions on a phone or laptop—for free, they can tolerate advertising, or pay a monthly subscription fee. In the west, that fee is considerably less than many music fans used to pay for individual CDs, cassettes, and vinyl, especially when reductions for families, students, and so on are taken into account. In the Global South, subscription streaming appeals to the growing middle classes in China, India, and Latin America.

Supermarkets and the corporations that dominate food production form partnerships that give them enormous power. The same is true of music platforms and the owners of the most popular music—namely, the "major" recording and publishing companies and the larger independents. And just as supermarkets can only combine low prices with profit by forcing down wages at suppliers, the limited amount paid by consumers for streaming inevitably limits what musicians can receive. While it has always been the case that most musicians can't make a living from recorded music, platforms have entrenched musical *cheapness* as a value. As Onur Sesigür's chapter in this volume shows, this has significant implications for musicians, such as those in Istanbul who turn to the Turkish advertising industry for work, with ambivalent consequences for their autonomy. And as Emília Barna illuminates in her chapter, in a semi-peripheral country such as Hungary, the conditions under which musicians work are strongly shaped by platforms' geographical policies, including playlists, the position of distributors on the global market, and deals between platforms and collecting societies—all marked by unequal power relations.

The dominance of supermarkets doesn't mean that food and drink have become less diverse. While the supermarket has spread globally, the form it takes and what people do with it vary. Food can still be obtained in many ways: via restaurants, cafes, traditional food markets, and contemporary farmers' markets, and some people grow their own vegetables and rear animals, even in the highly urbanized west. Similarly, the pleasures of shopping for new and secondhand records, CDs, and cassettes are still available, and the retail site Bandcamp offers some kind of online alternative. People still consume and perform live music at festivals, venues, and bars. A great deal of music is still enjoyed on radio and television.

Obviously, there are differences. Supermarkets differentiate by price and quality—Whole Foods and Waitrose versus Walmart and Aldi—whereas platforms offer pretty much the same repertoire and price as one another. Supermarkets have a highly visible offline presence, whereas most of us only ever experience MSPs online.[91] MSPs are personalized in a way that offline supermarkets can never be.

Yet the analogy is potentially enlightening. In both domains, a dominant sameness closes down alternatives, making it difficult to imagine other ways of doing things and portraying alternative forms of consumption as *inconvenient*—an option that requires a high level of ethical commitment, likely to be practiced by only a small number of consumers. At the same time, it would be wrong to see the spread of streaming as homogenization at the level of music itself. Rather, streaming represents an amplification, in the realm of culture, of the problematic abundance already fostered by modernization and industrialization; generative AI, with its hugely damaging environmental consequences, only adds to that destructive profusion. While access to the abundance offered by streaming is very unequally distributed, it would be simplistic to claim that it merely masks homogeneity, given the international mix of sounds available to audiences on streaming platforms. Instead, if there is homogenization, it is evident more in the way that MSPs, like supermarkets, embody western notions of flourishing via abundance and convenience, the latter exemplifying western notions of time as a linear resource that must be maximized. This is seen in its hypermodern form of time-space compression, whereby, in David Harvey's words, accelerating turnover time in production is linked with "parallel accelerations in exchange and consumption."[92] The collection and parsing of data about music and its uses is a key driver of these circuits of acceleration.

Then, of course, there is the problem of inequality, not only among consumers, but also among producers. As with supermarkets, consumption convenience for some goes hand in hand with worker exploitation and alienation. As digitalization emerged, some predicted a brighter future where increasing numbers of artists and smaller companies would be able to succeed. But for all their abundance and internationalization, the music industries still operate on a "winner take all" basis, as they have throughout their history: the most successful tracks and artists, and the biggest rights-holders (i.e., recording and publishing companies), dominate streams and therefore payments. Relatedly, while there were widespread concerns and controversies about musician remuneration from the early days of streaming, it has always been the case that most musicians have existed as precarious and underpaid cultural workers (though controversies about streaming have helped to bring about a new consciousness of these conditions in many places).

7   MULTIPOLARITY AND INEQUALITY

In this introductory chapter, I have summarized some ways that prevailing forms of musical production and consumption are evolving, examining the role of those technologies and business models apparent in discourses of "streaming." I have emphasized that streaming is associated with a developing musical multipolarity, defying predictions of homogenization that have characterized many jeremiads about culture and music ever since their industrialization began in earnest in the nineteenth century. Streaming undoubtedly offers convenient and relatively afford-able access to a remarkable abundance of music for hundreds of millions of people. There is plausible evidence that it is associated with the new global popularity of musics from outside the Anglo-American core that once dominated the international music industries (though whether music streaming has *brought about* that popularity is another matter altogether). Moreover, the music industries in the age of streaming are marked by an unprecedented complexity and, some would argue, diversity of industrial and organizational forms, including the possibility of reaching audiences more directly than was previously the case, as many of the chapters in this volume demonstrate. But as this book also shows, for all its growing multipolarity and apparent diversity, music in the age of streaming remains embedded in problematic assemblages of capital, colonialism, technology, and everyday life. Intellectual property and individualistic consumerism are fundamental features of these apparatuses, now supplemented by new dynamics of datafication, automation, and the power of digital platforms. The varied contributions to this collection navigate these interacting currents of multipolarity and inequality.[93]

NOTES

1. Some of the sense of loss often expressed is summarized by Kyle Devine, *Decomposed: The Political Ecology of Music* (Cambridge, MA: MIT Press, 2021), 9–10, though he refuses to engage with nostalgia, for or against; the purpose of his book is to investigate the environmentally damaging consequences of recorded music, including streaming's underlying materiality as well as older formats such as vinyl.

2. For a concise overview of platformization in general, see Thomas Poell, David Nieborg, and José van Dijck, "Platformisation," *Internet Policy Review* 8, no. 4 (2019), https://doi.org/10.14763/2019.4.1425;

and for a very widely cited treatment of platformization in the realm of culture, see Thomas Poell, David Nieborg, and Brooke Ann Duffy, *Platforms and Cultural Production* (Cambridge: Polity Books, 2022). Although Gillespie is dealing with social media and does not discuss music, he provides a way through the thicket of defining digital platforms in his *Custodians of the Internet* (New Haven, CT: Yale University Press, 2018), 17–21, recording various reservations, qualifications, and competing uses, while emphasizing that platforms (a) host, organize, and circulate content without having produced it; (b) rely on collecting and processing data for customer service and (often) advertising and profit; and (c) have to engage in content moderation (which is his main theme).

3. There is now an entire field of critical data studies. On datafication in general, see Ulises A. Mejias and Nick Couldry, "Datafication," *Internet Policy Review* 8, no. 4 (2019), https://doi .org/10.14763/2019.4.1428. On datafication and data capture in relation to music, see Robert Prey, "Musica Analytica: The Datafication of Listening," in *Networked Music Cultures*, ed. Raphael Nowak and Andrew Whelan (London: Palgrave Macmillan, 2016), 31–48; and Leslie M. Meier and Vincent R. Manzerolle, "Rising Tides? Data Capture, Platform Accumulation, and New Monopolies in the Digital Music Economy," *New Media and Society* 21, no. 3 (2019): 543.

4. There is also now an entire field of critical algorithm studies. A good overview of some key issues, including racial "bias," is Robyn Caplan, Joan Donovan, Lauren Hanson, and Jeanna Matthews, *Algorithmic Accountability: A Primer* (New York: Data and Society, 2018), https://datasociety.net/wp -content/uploads/2019/09/DandS_Algorithmic_Accountability.pdf.

On automated recommendation in music, see David Hesmondhalgh, Raquel Campos Valverde, D. Bondy Valdovinos Kaye, and Zhongwei (Mabu) Li, "The Impact of Algorithmically Driven Recommendation Systems on Music Consumption and Production: A Literature Review," UK Centre for Data Ethics and Innovation Reports, 2023, https://ssrn.com/abstract=4365916. An accomplished anthropology of how music recommendation systems developed is Nick Seaver, *Computing Taste: Algorithms and the Makers of Music Recommendation* (Chicago: University of Chicago Press, 2022).

5. David Hesmondhalgh and Leslie Meier, "What the Digitalisation of Music Tells Us about Capitalism, Culture and the Power of the Information Technology Sector," *Information, Communication and Society* 21, no. 11 (2018), 1555–70.

6. Terminology concerning international inequalities in wealth and power is haunted by dualisms: developed and developing, West and non-West, Global North and Global South, Minority World and Majority World. The first of these pairs has now been abandoned by many because of its colonialist connotations—that the "developing" nations should become like the "developed" rather than pursuing their own paths. Applied in a binary way, the other pairings risk downplaying the wealth and privilege of elites in the latter entities. North and South, East and West, are geographically problematic, not least given the economic and political power of China and Japan. For all these problems, critics of injustice need distinctions that capture inequalities in a divided world. The Majority/Minority World pairing—often attributed to an essay by Shahidul Alam, "Majority World: Challenging the West's Rhetoric of Democracy," *Amerasia Journal* 34, no. 1 (2008): 89–98—helpfully draws attention to the fact that most of the world's population, and implicitly most of the world's poor, are people of color. In the United Kingdom, the term "people of the Global Majority" has been increasingly adopted by activists seeking an alternative to bureaucratic nomenclature such as "ethnic minority" and "BAME" (Black, Asian, and Minority Ethnic). Nevertheless, in this chapter, I retain the terms "western" and "Global South," as I have been advised that they are more familiar to likely readers, including those with origins in poorer countries. I also employ terms such as "core" and "periphery," and "wealthier" and "less wealthy," partly because they allow for questions of degree (e.g., the possibility of describing some places as "semiperipheral"). Terms such as "western music streaming platforms" and "western music industries" here mainly refer to those with their origins in Western Europe, North America, and Japan. I also use "Euro-American" despite its potential exclusion of relatively wealthy countries such as Japan, Australia, and New Zealand. Thanks to my colleague Dibya Roy for a helpful exchange about "Majority World."

7. An academic monograph based on research conducted in response to public controversies about streaming was Maria Eriksson, Rasmus Fleischer, Anna Johansson, Pelle Snickars, and Patrick Vonderau, *Spotify Teardown: Inside the Black Box of Streaming Music* (Cambridge, MA: MIT Press, 2018). This was an important intervention, but its emphasis on Spotify is apparent in many other studies and debates, which would benefit from greater attention to the whole platform ecosystem surrounding music. A more recent book-length study is Tiziano Bonini and Paolo Magaudda, *Platformed! How Streaming, Algorithms and Artificial Intelligence Are Shaping Music Cultures* (Cham, Switzerland: Palgrave Macmillan, 2024).

8. On "surveillance" issues in relation to music, a notable contribution was Eric Drott's "Music as a Technology of Surveillance," *Journal of the Society for American Music* 12, no. 3 (2018): 233–67, reproduced in somewhat modified form as a chapter in Eric Drott, *Streaming Music, Streaming Capital* (Durham, NC: Duke University Press, 2024). Drott does not address wider debates about privacy and surveillance in digital networks, such as those considered in Julie Cohen, *Configuring the Networked Self* (Oxford: Oxford University Press, 2012).

9. Lanre Bakare, "The Music Streaming Debate: What the Artists, Songwriters and Industry Insiders Say," *Guardian*, April 10, 2021, www.theguardian.com/music/2021/apr/10/music-streaming -debate-what-songwriter-artist-and-industry-insider-say-publication-parliamentary-report; David Hesmondhalgh and Hyojung Sun, "How the Working Conditions of Musicians (Finally) Became a Matter of Mainstream Political Interest," in *Handbook of Critical Music Industry Studies*, ed. David Arditi and Ryan Nolan (New York: Palgrave, 2024), 605–25.

10. "Several submissions" to a UK parliamentary inquiry "warned that algorithms, as with any recommendation system, could reflect biases that may subsequently reduce new music discovery, homogenise taste and disempower self-releasing artists." Digital, Culture, Media, and Sport Committee, *Economics of Music Streaming* (London: House of Commons, 2021), 79, archived July 29, 2021, at https://web.archive.org/web/20210729114849/https://committees.parliament.uk/publications/6739 /documents/72525/default/.

11. With apologies for another self-citation, these controversies are summarized and discussed in David Hesmondhalgh, "Streaming's Effects on Music Culture: Old Anxieties and New Simplifications," *Cultural Sociology* 16, no. 1 (2022): 3–24. For a thoughtful essay on individualization via digital technologies, including streaming, see Nancy W. Hanrahan, "Digitized Music and the Aesthetic Experience of Difference," in *The Dialectic of Digital Culture*, ed. David Arditi and Jennifer Miller (Lanham, MD: Lexington Books, 2019), 165–76.

12. Some of the contributions to debates about the effects of streaming represent efforts to understand and evaluate baffling changes, in a search for more precise and valid critique; some tend toward simplification and even distortion and reproduce tired tropes of twentieth-century mass culture. See Hesmondhalgh, "Streaming's Effects." Given that music has indeed continued to mutate, as it always will, understanding the extent to which these and other developments truly exist—and if they do, whether such developments can be attributed to the rise of MSPs—is challenging.

13. I should emphasize that the perspective in this introductory chapter is not necessarily shared by authors of other chapters. My thanks to Georgina Born, Sumanth Gopinath, Toussaint Nothias, Anamik Saha, and Anjali Vats for their very helpful comments on a draft. Remaining faults are, of course, my responsibility.

14. I cannot do justice here to the work of scholars from the Global South and their Minority World allies who have struggled against the western- and Euro-American–centrism that have afflicted humanities research and education for so long. The notion of "decolonization" has become important in such struggles, supplementing and at times displacing "de-westernization." There are hundreds of publications applying these concepts to music studies, media studies, and internet studies, as well as dozens more debating how the terms should and should not be used. On the importance of the concept of decolonization and some limitations in how it has been applied, from an anticolonial perspective, see Leon Moosavi, "The Decolonial Bandwagon and the Dangers of Intellectual Decolonisation," *International Review of Sociology* 30, no. 2 (2020): 332–54.

15. One impressive contribution to music studies in this respect, which pays significant attention to digitalization, is Georgina Born, ed., *Music and Digital Media: A Planetary Anthropology* (London: UCL Press, 2022), which has essays featuring analysis of Argentina, Cuba, India, Kenya, North America, and Europe. Its valuable case studies are based on research preceding the global spread of music platformization.

16. Kofi Agawu, "Tonality as a Colonizing Force in Africa," in *Audible Empire: Music, Global Politics, Critique*, ed. Ronald Radano and Tejumola Olaniyan (Durham, NC: Duke University Press, 2016), 335. See also Nicholas Cook, "Western Music as World Music," in *The Cambridge History of World Music*, ed. Philip Bohlman (Cambridge: Cambridge University Press, 2013), 75–100, on how western music spread into Asia, Africa, and elsewhere from the nineteenth century onward. I understand colonialism as a "practice of domination, which involves the subjugation of one people to another," to quote Margaret Kohn, "Colonialism," *Stanford Encyclopedia of Philosophy*, last modified January 17, 2023, https://plato.stanford.edu/entries/colonialism/. In this chapter, I am referring mainly to European colonialism, which involved "European settlement, violent dispossession and political domination" over much of the rest of the world. Kohn points to the difficulties caused by the fact that the word is often used as a synonym for imperialism, which "often describes cases in which a foreign government administers a territory without significant settlement" (such as the late nineteenth-century "scramble for Africa" or US domination of the Philippines) and in which control might be more indirect, but that still very frequently involves dispossession and violence.

17. Agawu, "Tonality," 335–37.

18. Michael Denning, "Decolonizing the Ear: The Transcolonial Reverberations of Vernacular," in Radano and Olaniyan, *Audible Empire*, 35. In a powerful intervention, Dylan Robinson has discussed the formation, in the context of the aftermath of violent subjugation of North American Indigenous populations, of what he frames as particular sets of "listening positionality" associated with settler colonialism; see *Hungry Listening: Resonant Theory for Indigenous Sound Studies* (Minneapolis: University of Minnesota Press, 2020).

19. The leading history of the international recording industry is Pekka Gronow and Ilpo Saunio, *An International History of the Recording Industry*, trans. Christopher Moseley (London: Cassell, 1999). See also *The Continuum Encyclopedia of Popular Music of the World*, vol. 1, *Media, Industry and Society*, ed. John Shepherd, David Horn, Dave Laing, Paul Oliver, and Peter Wicke (London: Continuum, 2003); and Lee Marshall, ed., *The International Recording Industries* (Abingdon: Routledge, 2013).

20. Dave Laing, "The Recording Industry in the Twentieth Century," in Marshall, *International Recording Industries*, 33; and Gronow and Saunio, *International History*, 11–12.

21. See Peter Drahos with John Braithwaite, *Information Feudalism: Who Owns the Knowledge Economy?* (London: Earthscan, 2002), 74–79; and for a detailed history of Berne and its revisions, Richard Osborne, *Owning the Masters: A History of Sound Recording Copyright* (New York: Bloomsbury, 2023).

22. Drahos with Braithwaite, *Information Feudalism*, 76.

23. Devine, *Decomposed*, 54–63.

24. See Osborne, *Owning the Masters*, 49–77. "Neighboring," "performing," or "related" rights (the former term has different meanings in different legislatures) allow record companies and performers to be compensated for the public performance and broadcasting of their music, as well as to earn revenue from "synchronization" of music with film and television soundtracks. Revenue from these sources is collected and distributed via complex and often opaque networks of collection societies and other intermediaries. See Chris Cooke, *Dissecting the Digital Dollar*, 3rd ed. (London: Music Managers Forum, 2020) for a good guide to the complexities of music copyright, though mainly referring to the United Kingdom and the United States.

25. Michael Frishkopf, "Nationalism, Nationalization and the Egyptian Music Industry," *Asian Music* 39, no. 2 (2008): 28–58; Yiu-Wai Chu, *Hong Kong Cantopop: A Concise History* (Hong Kong: HKU Press, 2017).

26. Such musics have long been a matter of fascination for ethnomusicologists and popular music studies scholars. There are literally thousands of relevant studies. For an ambitious early survey, see Peter Manuel, *Popular Musics of the Non-Western World* (Oxford: Oxford University Press, 1988).

27. On the impact of Jamaican reggae on Euro-American popular song, see Michael E. Veal's *Dub: Soundscapes and Shattered Songs in Jamaican Reggae* (Middletown, CT: Wesleyan University Press, 2007). On hip hop, see Marcyliena Morgan and Dionne Bennett, "Hip-Hop and the Global Imprint of a Black Cultural Form," *Daedalus* 140, no. 2 (2011): 176–96.

28. Michael Denning, *Noise Uprising: The Audiopolitics of a World Musical Revolution* (London: Verso, 2015), 38.

29. Denning, "Decolonizing the Ear," 30, italics in the original.

30. Sabrina Petra Ramet, ed., *Rocking the State: Rock Music and Politics in Eastern Europe and Russia* (New York: Routledge, 1994).

31. To give just one example, the resilience of sacred musics such as kirtan and qawwali in South India and its diasporas continues to this day. See Virinder S. Kalra, *Sacred and Secular Musics: A Postcolonial Approach* (London: Bloomsbury, 2016).

32. Gronow and Saunio, *International History*, 183. See also Peter Manuel, *Cassette Culture: Popular Music and Technology in North India* (Chicago: University of Chicago Press, 1988).

33. Ruth L. Gana, "Has Creativity Died in the Third World? Some Implications of the Internationalization of Intellectual Property," *Denver Journal of International Law and Policy* 24, no. 1 (1995): 109–44.

34. Marley's rise was dependent on the London-based postcolonial record industry. See Jason Toynbee, *Bob Marley: Herald of a Post-Colonial World* (Cambridge: Polity Books, 2009).

35. Jeremy Wade Morris, *Selling Digital Music, Formatting Culture* (Berkeley: University of California Press, 2015).

36. Jonas Andersson Schwarz, *Online File Sharing: Innovations in Media Consumption* (London: Routledge, 2013).

37. Shoshana Zuboff, *The Age of Surveillance Capitalism* (London: Profile, 2018).

38. Julie Cohen, *Between Truth and Power: The Legal Constructions of Information Capitalism* (Oxford: Oxford University Press, 2019).

39. In some parts of the world, phone companies were key to the early digital industry and retain dominance in some places, such as Gaana in India and Boomplay in Nigeria. See Christine Ithurbide, "Telecom and Technology Actors Repositioning Music Streaming," in *Platform Capitalism in India*, ed. Adrian Athique and Vibod Parthasarathi (Cham, Switzerland: Springer, 2020), 107–24.

40. Blake Durham and Georgina Born, "Online Music Consumption and the Formalisation of Informality: Exchange, Labour and Sociality in Two Music Platforms," in Born, *Music and Digital Media*, 177–219.

41. For accounts emphasizing these infrastructural dimensions of change, see David Hesmondhalgh, Raquel Campos Valverde, D. Bondy Valdovinos Kaye, and Zhongwei (Mabu) Li, "Digital Platforms in the Realm of Culture," *Media and Communication* 11, no. 2 (2023): 296–306; and Zhongwei (Mabu) Li and David Hesmondhalgh, "From P2P to the Cloud: Music, Platformization and Infrastructural Change in China," *Chinese Journal of Communication* (2024): 1–17. See also Li and Kaye's chapter in this volume (chapter 10).

42. The terms of their music licensing agreements with music rights-owners meant that streaming services were not allowed to produce their own musical content. This is one of the reasons why Spotify has invested heavily in podcasts. The major rights-holders, plus a consortium of larger independents (Merlin), even had financial stakes in Spotify, though the size and consequences of these stakes have often been overstated by critics (at the time of writing, Universal and Sony have roughly 3 percent each).

43. Google opened its first "full-service" music streaming platform in 2013 and acquired Songza in 2014 to improve its recommendation systems; its service currently operates as YouTube Music. Apple acquired Beats Music (and Beats Electronics) in 2014 and launched its Apple Music streaming service in

2015. Amazon launched its full streaming service, Amazon Music, in 2016. By 2024, the MSPs operated by Amazon, Apple, and YouTube accounted for about 32 percent of global subscriptions, with Spotify achieving around the same amount. Mark Mulligan, "Music Subscriber Market Shares: Slowdown? What Slowdown?," MIDiA Research, March 27, 2025, www.midiaresearch.com/blog/music-subscriber-market-shares-2024-slowdown-what-slowdown.

44. According to Ellis Jones, digital platforms offer DIY musicians such as himself, "in return for our circulatory zeal," a set of tools, "a wealth of accessible, automated solutions"—though Jones also provides an excellent critical account of the way that social media platforms bring about a depoliticization of DIY production aesthetics. See Jones, *DIY Music and the Politics of Social Media* (London: Bloomsbury, 2021), 121–36.

45. Prominent independent distributors internationally include Amuse and DistroKid, but also companies such as AWAL and The Orchard, both of which were eventually purchased by Sony.

46. See Daniel Tencer, "158 Million Tracks Had 1,000 Plays or Fewer on Music Streaming Services Last Year. 45 Million Had No Plays at All," *Music Business Worldwide*, January 10, 2024, www.musicbusinessworldwide.com/158-million-tracks-1000-plays-on-streaming-services/.

47. See David Hesmondhalgh, Hyojung Sun, Richard Osborne, and Kenny Barr, *Music Creators' Earnings in the Digital Era* (London: Intellectual Property Office, 2021), 194–215.

48. Bethany Klein, *Selling Out: Culture, Commerce and Popular Music* (London: Bloomsbury, 2020).

49. Ana Alacovska, "'Keep Hoping, Keep Going': Towards a Hopeful Sociology of Creative Work," *Sociological Review* 67, no. 5 (2019): 1118–36.

50. For a discussion of this homogeneity and the potential significance of these categories for understanding change and continuity in musical experience, see David Hesmondhalgh, Raquel Campos Valverde, D. Bondy Valdovinos Kaye, and Zhongwei (Mabu) Li, "Critically Analyzing Platform Interfaces: How Music Streaming Platforms Frame Musical Experience," *International Journal of Communication* 18 (2024): 3257–80. On debates about function and mood more broadly, see Hesmondhalgh, "Streaming's Effects."

51. "Country Profile—Mexico 2023," *Music Ally*, November 7, 2023, https://musically.com/2023/11/07/country-profile-mexico-2023/.

52. Robert Prey and Seonok Lee, "A Global Approach to Studying Platforms and Cultural Production," in "Global Perspectives on Platforms and Cultural Production," special issue, *International Journal of Cultural Studies* 28, no. 1 (2024), https://doi.org/10.1177/13678779241244399.

53. Prey and Lee, "Global Approach."

54. I offer a history of these developments across multiple cultural sectors in David Hesmondhalgh, *The Cultural Industries*, 5th ed. (London: Sage, 2025).

55. Most apparent perhaps in the work of communication scholar Daniel Lerner, *The Passing of Traditional Society: Modernizing the Middle East* (New York: Free Press, 1958).

56. Annabelle Sreberny-Mohammadi, "The Many Cultural Faces of Imperialism," in *Beyond Cultural Imperialism: Globalization, Communication and the New International Order*, ed. Peter Golding and Phil Harris (London: Sage, 1997), 49. A frequently cited article by Eve Tuck and K. Wayne Yang, "Decolonization Is Not a Metaphor," *Decolonization: Indigeneity, Education and Society* 1, no. 1 (2012): 1–40, warns of the dangers of using anticolonialist terms metaphorically, including "colonialism" (and, by extension, "imperialism"). Others argue that such an approach might downplay how the term "decolonization" might help expose and counter the myriad ramifications of colonialism beyond land capture and economic exploitation; see Gurminder K. Bhambra, Dalia Gebrial, and Kerem Nişancıoğlu, "Introduction: Decolonising the University," in *Decolonising the University*, ed. Bhambra, Gebrial, and Nişancıoğlu (London: Pluto, 2018), 5. See also note 16 above, on potential confusions about the differences or otherwise between "colonialism" and "imperialism."

57. See Hesmondhalgh, *Cultural Industries*, 376, where I draw on my own earlier efforts to synthesize and assess research on cultural imperialism, for example, Chin-Chuan Lee, *Media Imperialism Reconsidered: The Homogenizing of Television Culture* (Beverly Hills, CA: Sage, 1980).

58. The locus classicus is the work of Alan Lomax, but related concerns were expressed by many ethnomusicologists. Sometimes these concerns were framed in ways that involved simplifications of the cultures and communities that produced them.

59. These debates are reflected in many contributions to Georgina Born and David Hesmondhalgh, eds., *Western Music and Its Others: Difference, Appropriation, and Representation in Music* (Berkeley: University of California Press, 2000).

60. Among a number of significant studies, see Aswin Punathambekar, *From Bombay to Bollywood* (New York: NYU Press, 2013).

61. Some of the work on hybridity (which is related to preceding concepts for understanding cross-cultural flows such as creolization and syncretism) was in danger of understating power inequalities. For a fine treatment that retains a focus on structural power, see Marwan Kraidy, *Hybridity, or the Cultural Logic of Globalization* (Philadelphia: Temple University Press, 2005).

62. Among important contributions specifically on music in relation to themes associated with cultural imperialism, see Dave Laing, "The Music Industry and the Cultural Imperialism Thesis," *Media, Culture and Society* 8, no. 3 (1985): 331–41; Simon Frith, "Anglo-America and Its Discontents," *Cultural Studies* 5, no. 3 (1991): 263–69; and Martin Stokes, "Music and the Global Order," *Annual Review of Anthropology* 33 (2004): 47–72.

63. Marc Raboy and Normand Landry, *Civil Society, Communication and Global Governance* (New York: Peter Lang, 2005).

64. Dal Yong Jin, *Digital Platforms, Imperialism and Political Culture* (London: Routledge, 2016); Michael Kwet, "Digital Colonialism: US Empire and the New Imperialism in the Global South," *Race and Class* 60, no. 4 (2019): 3–26; Paola Ricaurte, "Data Epistemologies, the Coloniality of Power and Resistance," *Television and New Media* 20, no. 4 (2019): 350–65.

65. For a good survey, see Nick Couldry and Ulises Ali Mejias, "The Decolonial Turn in Data and Technology Research: What Is at Stake and Where Is It Heading?," *Information, Communication and Society* 26, no. 4 (2023): 786–802.

66. A more promising source of treatments of international cultural inequality in the digital era (including some reference to music) comes from the subfield of media and communication studies that addresses "platform studies." See Philippe Bouquillion, Christine Ithurbide, and Tristan Mattelart, *Digital Platforms in the Global South: Reconfiguring Power Relations in the Cultural Industries* (London: Routledge, 2024); and Thomas Poell et al., "Global Perspectives on Platforms and Cultural Production," *International Journal of Cultural Studies* 28, no. 1 (2024), https://doi.org/10.1177/13678779241292736, an introduction to "Global Perspectives on Platforms and Cultural Production," a special issue of twenty articles.

67. An exception is Sara Bannerman, "Platform Imperialism, Communications Law and Relational Sovereignty," *New Media and Society* 26, no. 4 (2024): 1816–33, though this does not significantly engage with theorization of imperialism and colonialism. Intriguingly, one of the key books in this tradition of IP critique uses the concept of feudalism (rather than capitalism or colonialism) as a lens: Drahos with Braithwaite, *Information Feudalism*.

68. For just one significant contribution, see Margaret Chon, "Intellectual Property and the Digital Divide," *Cardozo Law Review* 27, no. 6 (2006): 2821–2912.

69. Significant exceptions include an exciting study that combined critical legal studies with cultural studies and anticolonial theory: Rosemary J. Coombe, *The Cultural Life of Intellectual Properties* (Durham, NC: Duke University Press, 1998).

70. An exception would be work on how western publishing corporations, using copyright to maximize profits in the Global South, actively prevented the affordable availability of educational materials such as textbooks. See Drahos with Braithwaite, *Information Feudalism*, 74–79. Margaret Chon rightly observes "an over-focus on entertainment products such as music and movies" in US copyright scholarship of the globalization era (Chon, "Intellectual Property," 2886), but little of the work that referred to entertainment came from a perspective critical of colonialism or international inequality.

71. For example, Nairobi activist Nanjala Nyabola's Kiswahili Digital Rights Project. A valuable overview of this and other important initiatives is provided by Toussaint Nothias, "How to Fight Digital Colonialism," *Boston Review*, November 14, 2022, www.bostonreview.net/articles/how-to-fight-digital -colonialism/.

72. See Justin O'Connor, *Culture Is Not an Industry* (Manchester: Manchester University Press, 2024); and David Hesmondhalgh, *Why Music Matters* (Malden, MA: Wiley, 2013).

73. Dwayne Winseck, "The Geopolitical Economy of the Global Internet Infrastructure," *Journal of Information Policy* 7 (2017): 228–67.

74. Spotify Technology S.A., "Form 10-K for the Fiscal Year Ended December 31, 2023," SEC filing, February 8, 2024, https://investors.spotify.com/financials/sec-filings-details/default.aspx ?FilingId=17248755; Apple Inc., "Apple Services Now Available in More Countries around the World," press release, April 28, 2020, www.apple.com/newsroom/2020/04/apple-services-now-available-in -more-countries-around-the-world/. In some of these countries, take-up of streaming is very low, particularly where credit cards and electronic forms of banking are not widespread.

75. Samuel F. Way, Jean Garcia-Gathright, and Henriette Cramer, "Local Trends in Global Music Streaming," *Proceedings of the Fourteenth International AAAI Conference on Web and Social Media* 14, no. 1 (2020): 705–14.

76. Way, Gathright, and Cramer, "Local Trends," 706. This represents a reversal of a trend identified by Estrella Gómez-Herrera, Bernice Martens, and Joel Waldfogel, "What's Going On? Digitization and Global Music Trade Patterns since 2006," European Commission, Institute for Prospective Technological Studies (2014), https://joint-research-centre.ec.europa.eu/publications/whats-going-digitization -and-global-music-trade-patterns-2006_en.

77. Figures from the data company Luminate, cited in Murray Stassen, "English Language Music's Popularity on Streaming Services Is Shrinking in the US ... and Globally," *Music Business Worldwide*, July 13, 2023, www.musicbusinessworldwide.com/from-muberts-100m-ai-generated-tracks-to-luminates -midyear-music-report-its-mbws-weekly-round-up1/.

78. Will Page and Chris Dalla Riva, "'Glocalisation' of Music Streaming within and across Europe," LSE Europe in Question Discussion Paper 182 (May 2023), www.lse.ac.uk/european-institute/Assets /Documents/LEQS-Discussion-Papers/EIQPaper182.pdf.

79. Leo Morel, "Brazil's Music Boom: Streaming, Sertanejo, Anitta and More," *Music Ally*, November 28, 2018, https://musically.com/2018/11/20/brazil-music-streaming-sertanejo-anitta/.

80. Much of this music, and its musicians, are Afro-Latinx; its success builds on that of an earlier wave of popularity of Latin popular music in the 1990s and early 2000s; see Noel Allende-Goitía, "The Rise of Afrodiasporic Meta-Genres and Global Afro-Latinx Music," in *Routledge Handbook of Afro-Latin American Studies*, ed. Bernd Reiter and John Antón Sánchez (New York: Routledge, 2023), chap. 26. On language issues, see Christopher Joseph Westgate, "The First Spanish-Language Album to Reach Number One: *El Último Tour del Mundo*, Bad Bunny, and the Billboard 200," *International Journal of Communication* 18 (2024): 1724–44.

81. For a business studies perspective of its recent rise (building on an earlier wave), see Jimmyn Parc and Shin Dong Kim, "The Digital Transformation of the Korean Music Industry and the Global Emergence of K-Pop," *Sustainability* 12, no. 18 (2020): 7790.

82. Desiree Ibekwe, "African Music to the World," *New York Times*, May 25, 2024, www.nytimes .com/2024/05/25/briefing/african-music-to-the-world.html.

83. For a survey of the developing research on this topic, and a bibliography, see Travis Harris, "Can It Be Bigger Than Hip Hop? From Global Hip Hop Studies to Hip Hop," *Journal of Hip Hop Studies* 6, no. 2 (2019), https://doi.org/10.34718/27nk-bx98.

84. Spotify, "Nearly a Quarter of All Streams on Spotify Are Hip Hop," *For the Record*, August 10, 2023, https://newsroom.spotify.com/2023-08-10/hip-hop-50-murals-new-york-atlanta-miami -los-angeles/.

85. Denning, "Decolonizing the Ear," 30. Thanks to Sumanth Gopinath for directing me to this passage in Denning's essay.

86. Denning, "Decolonizing the Ear," 31.

87. Denning, "Decolonizing the Ear," 34–35.

88. Indeed, my experience is that these new globally circulating genres emanating from the Global South are often heard by western liberal outsiders as being "like hip hop," particularly in their performative displays of self-confidence and sexuality.

89. Thomas Reardon, Peter Timmer, Christopher B. Barrett, and Julio Berdegué, "The Rise of Supermarkets in Africa, Asia, and Latin America," *American Journal of Agricultural Economics* 85 (2003): 1140–46.

90. See Hesmondhalgh et al., "Critically Analysing Platform Interfaces."

91. For a brief period in the 1990s and early 2000s, "megastores" mirrored supermarkets in selling physical music artifacts. In this volume, Eisenberg recounts how supermarkets were the basis of one early experiment in how the Kenyan recording industry sought to deal with the challenge of "piracy" in the 1990s.

92. David Harvey, *The Condition of Postmodernity* (Oxford: Blackwell, 1989), 285.

93. The contributions to this volume lean more toward examining relations between industry and technology in the context of mutating forms of capitalism and colonialism. They do not discuss how music streaming technologies might be (re)shaping musical subjectivities—and how this relates to debates about other digital and social media. Shuwen Qu and I, along with other colleagues, are exploring this and related issues, such as the individualization of music consumption mentioned above, in our current work.

2

Platformization and the Recording Industry in Kenya

Andrew J. Eisenberg

Musical platformization, understood as a process of "infrastructural transformation" through which a music recording industry is reorganized around digital content platforms and music creators repositioned as "platform complementors,"[1] has been under way in Kenya for well over a decade now. For much of this time, the process could adequately be described as one of transectorial innovation involving an array of "dynamic alignments" between the mobile telecommunications and technology sector and the music sector.[2] But the recent entrance of the globally dominant streaming service Spotify and China-headquartered African powerhouse Boomplay into the Kenyan market signals the arrival of a new phase of musical platformization in Kenya centered on streaming platforms and their curated playlists.

Drawing on data from qualitative research carried out intermittently over more than a decade, this chapter reviews the history of musical platformization in Kenya, focusing on its impact on the local recording industry.[3] My twofold aim is to offer an empirically rich case study of African "musical capitalism"[4] in the age of digital platforms while also contributing to the growing literature exploring histories of digital platformization outside of the Global North.[5] Two insights emerge from my account. One concerns the relationship between platformization and piracy, a major concern of industry and scholarly discussions of musical platformization in the Global North. Whereas the overriding focus in work on musical platformization in the Global North has been its relationship to "online piracy,"[6] what follows suggests that musical platformization in Kenya was accelerated by the broader market failure underpinning all forms of unlicensed distribution of Kenyan popular music. The other insight concerns how the Kenyan recording industry has been

32

TABLE 2.1 Phases of musical platformization in Kenya

Phase	Approximate dates	Key format	Platforms	Aggregators
1	2004–2011	MP3	Websites	Individual agents
2	2011–2018	Caller ringback tone	Mobile value-added services	Digital content firms (aka "content providers" or "premium rate service providers")
2(a)	2011–2015	MP3	Experiments with web/mobile apps for music downloads and legal sharing	n/a
3	2018–	Stream	Music streaming services and social media platforms	Digital distribution companies

transformed by platformization. In short, my account reveals that the processes of "reintermediation"[7] that have accompanied musical platformization in Kenya have thoroughly *reconfigured* the local recording industry, redefining and reorganizing roles and relations well beyond just those of the new intermediaries themselves.

I proceed by offering a periodization of musical platformization in Kenya, delineating three broad phases (plus one interstitial phase), each centered on a different format of music distribution (see table 2.1). Viewed on its own, phase 1 was not really a process of platformization. However, it involved the introduction of incipient platforms and set the stage for the significant institutional and infrastructural transformations of phases 2 and 3.

PHASE 1

The simultaneous liberalization of the broadcast media and the advent of affordable digital music technologies in the late 1990s sparked the emergence of a new recording industry in Kenya characterized by small labels and independent producers, digital production techniques, and creative engagements with hip hop and dancehall styles. Centered in Nairobi, this new industry emerged alongside an already existing recording industry in the city's downtown River Road district, which continued to thrive on "vernacular" (ethnically exclusive) popular musics and Swahili-language gospel. For all this vitality in music production, phonogram distribution in Kenya was in poor shape. The rampant commercial piracy that had contributed to driving the multinational record companies out of Nairobi in the 1980s continued unabated. The system of distribution linked to the old recording industry in downtown River Road worked well enough, outcompeting piracy through sheer speed and efficiency in delivering the music to consumers of vernacular and gospel musics within and beyond Nairobi.[8] But this success proved extremely difficult to replicate for the new recording industry, which had to start from scratch with new audiences that were still taking shape.

The collision of commercial piracy and the renaissance of local music production in Kenya at the turn of the millennium set the stage for digital forms of music distribution and, by extension, musical platformization in the country. But with the internet still in its infancy in the 1990s, early agents of Kenya's new recording industry remained focused on making physical distribution work, despite the difficulty. One strategy that had some limited success for them was the use of supermarkets as music retail outlets. This approach, which took shape initially through a partnership between the Uchumi supermarket chain and music label/distributor Serenade Music, ended up playing a surprising role in what I am calling phase 1 of musical platformization in Kenya.

Between 2004 and 2006, shoppers at certain Uchumi locations could purchase Kenyan popular music recordings not only in the form of professionally packaged cassettes and CDs distributed by Serenade Music, but also as personalized CD compilations available from a kiosk run by MyMusic, a company founded by Kenyan music manager Fakii Liwali with the assistance of Kenyan software engineer Bernard Kioko. Contrary to how it may have appeared to Uchumi shoppers, MyMusic was a digital business, centered on a website designed and maintained by Kioko's company, Bernsoft. MyMusic.co.ke was a marketplace for downloadable MP3s of Kenyan popular music singles and albums, geared primarily toward credit card–holding Kenyans abroad, especially middle-class university students in places such as South Africa and the United States, who were hungry to stay in touch with the fast-changing youth culture back home. In the manner of what would now be called a "platform," the site had a backend API that provided sales numbers to rightsholders in addition to a digital storefront.

The supermarket distribution model quickly proved unsustainable for MyMusic due to the personnel and equipment it required, as well as the cash flow problems associated with relying on Uchumi to run receipts. The website held more promise, especially with the advent of the mobile money system M-Pesa in early 2007, which enabled MyMusic to sell to the vast majority of Kenyans without credit cards. Nevertheless, MyMusic.co.ke ceased operations less than a year after the end of the deal with Uchumi. In Liwali's estimation, this failure was not due to the business model itself. The problem, he told me, was that Kenyan music labels and artists at the time failed to appreciate "the importance of speed" in competing with commercial piracy at home and unlicensed file sharing abroad: "You'd find that someone brings you a song to upload onto the system, and by the time they bring it, it's maybe two to four weeks into the market already." In other words, MyMusic's specific failure was its failure to persuade other parties to conform to the temporality of informal distribution.[9]

Arguably, MyMusic suffered from being ahead of the curve. Similar websites that emerged in its wake did better, perhaps benefiting from MyMusic's efforts to enroll Kenyan labels and artists in the new network of relations that was music e-commerce. These sites included the interconnected KenyanGospel.com and KenyanDownloads.com (2006–13), KenTunes.com (2009–17), and PewaHewa. com (2010–14). Along with selling music, all of them followed the lead of MyMusic in offering some free music downloads and streams, as well as an array of other free content, ranging from exclusive news stories to images (e.g., "wallpapers") and concert videos. The supplemental content was mainly geared toward members of "the Diaspora," as Kenyans living abroad like to call themselves, providing opportunities for them to stay informed and connected to the fast-moving developments in popular culture back home. In this Diaspora focus, the sites followed the examples of two earlier websites established by Diaspora Kenyans, which provided MP3 downloads of Kenyan popular music for free (often with the express permission of the artists, who saw the Diaspora as a source of lucrative performance possibilities), but otherwise had all the features of MyMusic and the others: a blog called KenyanMadness.8k.com (2001–5), established by Kenyan Christian hip hop artist Richard "Astar" Njau while he was at university in South Africa, and the web magazine Mwafrika.com (2006–12), established by Kenyan Christian music manager and event organizer David Kuria upon his return from university in Australia. These men soon went on to participate in the formal e-commerce industry: Njau became the local licensing agent for KenTunes.com, whose proprietor, Kevin Muthuri, was based in the United States. Later, both men worked together on PewaHewa.com, which Kuria founded.[10]

Phase 1 of musical platformization brought limited changes to Kenya's recording industry compared to subsequent phases. But the shifts it did introduce were similar in kind, if not in scope, to those introduced by phases 2 and 3. Working as small-scale (we might say "artisanal") content aggregators, agents like David Kuria and Richard Njau engaged with music and music creators in ways that subtly modulated existing understandings of professional roles and commercial products within the industry.

PHASE 2

In 2009, Kenya's largest mobile network provider, Safaricom, launched Skiza Tunes, a value-added service (VAS) offering "caller ringback tones" (music or other audio content to replace the standard signal tone that a caller would normally hear when dialing the customer) in exchange for a small subscription fee of around US$0.01 per day. With an annual gross revenue in excess of US$57 million per year (per 2022 statistics), Skiza can be viewed as a force of "disruptive innovation" for the Kenyan recording industry in and of itself.[11] But as I have described elsewhere, it is also the centerpiece of a broader convergence of mobile telecommunications and music in Kenya.[12] This is what I am referring to as phase 2 of musical platformization in Kenya.

The advent of phase 2 coincided with an explosion of research, entrepreneurship, and investment in digital applications in Kenya. In November 2010, Kenya's major daily, *Daily Nation*, reported that a combination of "innovations driven by mobile telephony and strong entrepreneurial spirit" had "created a magnet of venture capitalists in Kenya seeking to fund tech startups."[13] By 2011, *The Economist* declared that Nairobi was becoming "an African tech hub," leading the way in the establishment of tech "incubators," organizations that "provide start-ups with advice and cheap spaces to work, in exchange for a stake."[14] The remarkable success of the mobile phone–based money transfer service M-Pesa, developed by Kenyan mobile network operator Safaricom in partnership with Vodafone, was a major catalyst for all this activity and played a central role in earning Kenya's technology sector the nickname "Silicon Savannah."

Kenya's caller ringback tone market quickly came to dwarf the MP3 download market. It also quickly proved lucrative for artists, particularly gospel and "vernacular" musicians, whose "content" (and, in the case of gospel musicians, morally upright messaging) appealed to the Kenyans of lower economic classes, who fell in love with caller ringback tones. The drawback was that returns for music rightsholders were minuscule per unit compared to MP3 downloads. Indeed, the payout per ringback subscription on Skiza was shockingly small when compared to the payout per download on PewaHewa.com. The issue was raised in an interview in early 2012 by Nanjira Sambuli, who is now a noted policy analyst but at that time was a recent college graduate working as a singer and music manager. Reflecting on her experience working with Kenyan Afro-fusion star Eric Wainaina to craft a distribution and marketing strategy, she noted that while PewaHewa ended up being a good revenue earner for Wainaina, Skiza did not even seem worth trying.

Of course, a ringback tone is a fundamentally different product than a downloaded MP3. Just as Keith Negus notes with respect to a music "stream"—but far more obviously in the case of a musical product that a buyer (subscriber) merely uses to "define their personality"—a ringback tone is not "music" to a corporation involved in delivering it.[15] It is "a means to another end rather than an end in itself."[16] Sambuli and Wainaina certainly understood this point, which has always been plain to Kenyan music rightsholders. But they also understood something that has been equally obvious to most, if not all, Kenyan music rightsholders—that revenues for Skiza and other caller ringback tone services are depressed by the proliferation of intermediaries in the caller ringback tone industry and their oligonomic configuration.[17] As Sambuli noted:

> [Caller ringback tones] is where the cuts have a bigger issue because with PewaHewa the payments are split two ways—so PewaHewa get their cut, and the artists get their cut. As opposed to [caller ringback tones], where . . . of course, the service provider gets his cut, [as do] the distributors of the content (because you don't supply the music directly to like Safaricom or the service provider; it goes to a distributor).[18]

The percentage of revenues siphoned off by various players in the caller ring-back tone industry has been a common complaint among Kenya's music rights-holders, especially independent artists, through a decade of struggle over royalties. Nevertheless, just a few years after my conversation with Sambuli, Wainaina had become a defender of the intermediaries who were taking chunks out of his own royalties. Responding to the news of a court order directing Safaricom to disburse Skiza royalties directly to collective management organizations (CMOs, i.e., royalty collecting societies) instead of content service providers—a ruling that would later be overturned—Wainaina expressed doubt about the wisdom of "cutting off the middlemen":

> He discloses that they (middlemen) are instrumental in marketing artistes' songs, "In fact, they act like record labels because they pay for music videos which cost a pretty penny." "Do you know how much my Celina video cost? Sh800,000, and how much did I pay for it? Zero. They bring in world class video directors so when your song starts making money they'll take 50 percent of your revenue," explains Wainaina.[19]

Wainaina's embrace of Skiza "middlemen" demonstrates how the intermediaries involved in phase 2 had reconfigured the recording industry at the height of their power. Christiaan De Beukelaer and I explore this phenomenon in an article that places the Kenyan case alongside similar developments in Ghana in order to tell a larger story about changes to music economies across the African continent.[20] Our approach is to map "the *dynamic alignments* of the music and MTT sectors in Ghana and Kenya, by which we mean the sometimes fleeting, sometimes lasting situations in which the two sectors come to share a set of institutional structures and strategies."[21] In an earlier, preliminary report, I describe the situation in terms more resonant with Wainaina's comments, as a matter of mobile telecommunication and content firms taking on, and thereby magnifying, roles within the Kenyan recording industry that had been relatively absent since the 1980s.[22] These are complementary perspectives. But I want to bring to bear my initial analytical lens here because it provides for a sharper connection to the other phases of musical platformization. The remainder of this section draws from my 2012 paper.

Mobile Telecommunications and Technology Firms as Talent Managers

MTECH

MTech East Africa, a subsidiary of MTech Nigeria, was headed in 2012 by a dapper Nigerian businessman named Ikechukwu Anoke. Iyke, as he likes to be called, had received some media attention for his efforts to connect Kenyan artists to the Nigerian music industry and market. While he was portrayed in the press as a music magnate, he was, in reality, a mobile technology executive. MTech East Africa isn't a music label or management company but a "digital solutions" firm that, at the time, primarily functioned as a music "content provider"—an intermediary set up to license and aggregate music for mobile service providers. Even so, in Iyke's hands, MTech was fast becoming a talent management agency. My only interview with him ended when our lunch table filled up with famous Kenyan music artists who had come to hang out. One of these artists was dancehall singer Wyre, who, as I would later learn, had signed on to be managed by Iyke.

Iyke attributed MTech's move into talent management to an increasingly competitive business market. In light of Indian firms like Spice VAS entering the digital content market, he told me, content providers must have something special to offer to artists. For MTech, this was a connection to Nigeria.

Safaricom

Safaricom was also entering the talent management game with Kenya Live, a series of well-choreographed concerts across the country. These shows featured an array of established Kenyan artists representing genres ranging from gospel to hip hop to "vernacular," all performing with a live band. In addition to intensive rehearsals, the artists were put through an "academy" with master classes on everything from vocal techniques to makeup. According to one of its directors, the objective was to bring seasoned Kenyan music performers "to the next level."[23]

In its literature, Safaricom framed this training as a way of helping to develop an industry that supplies content for its highly profitable VAS platforms. In other words, Safaricom saw the program as part of its broader work in supporting the Kenyan music industry, which also included sponsoring the industry expo, Kenya Music Week. Each performer contracted for the Kenya Live academy and tour also agreed to work with the media house Homeboyz Entertainment to produce exclusive content for Safaricom to sell, including customized ringtones and wallpapers.

Mobile Telecommunications and Technology Firms as Labels

In marketing and selling this exclusive content, Safaricom was positioning itself as something of a minilabel. Another firm within the mobile phone sector that started a minilabel was MyMusic.co.ke's Bernsoft, which had become a digital content firm by this point. The company opened its own recording studio. To inaugurate the facility, Bernsoft founder and CEO Bernard Kioko commissioned a patriotic song for a nongovernmental, apolitical campaign for national unity. In doing so, he was able to get some of Kenya's most famous artists to participate in the pilot project free of charge.

PHASE 2(A)

Writing for the *East African* at the beginning of 2014, information and communications technology analyst Russell Southwood declared 2013 "the year that digital content in Africa began to become a mass reality," suggesting that "content delivery [in Africa] may see power shift from telcos."[24] His comments reflected the general sense among digital entrepreneurs in Kenya and other African countries that a boom in media streaming was on the horizon, and that the beneficiaries would be those who arrived early to the party. Hence, between 2012 and 2014, a bevy of African music streaming start-ups appeared, with more than one of them being touted as the "iTunes for Africa."

The most significant Kenyan start-up during this period was Mdundo, which is still going strong today. Mdundo emerged out of 88mph, a "tech seed fund and accelerator" founded by Danish entrepreneur Kresten Buch. Buch was one of the founders of the business, along with Martin Nielsen, who has served as CEO from the start, and Francis "Frasha" Amisi, a Kenyan hip hop artist. Nielsen explained that the initial inspiration for the project came from discussions about the failure of phonogram distribution in Kenya and the idea that the example of Spotify may provide the answer.[25]

True to the character of what I call phase 2(a) of musical platformization in Kenya, Mdundo's initial business model was highly experimental. It involved the use of "scratch card vouchers," similar to those used for selling mobile phone minutes, which artists could sell or give away to customers. Customers could use the cards to download songs, and then, as Martin Nielsen explained, Mdundo would take a split of purchases that followed the initial free downloads. Innovative as it was, this model "massively failed" because "no one bought the cards."[26] Mdundo ultimately switched to a freemium download model (offering ad-supported free downloads and a paid premium tier, where downloads are free of advertisements), which remains in place today.

Along with Mdundo, another digital music start-up in Kenya that received press attention in 2014 was Waabeh, an "audio marketplace" developed by a team led by Kenyan producer Tim Rimbui. Though it only lasted a few years, Waabeh started out strong with a deal to have the platform preinstalled on the new Yolo phone, an Intel device developed for Safaricom. Beneath the surface of Mdundo and Waabeh was an array of other initiatives for platforming Kenyan popular music, often involving fantastically creative approaches to solving the problem of distribution. In each case, the project was inspired by personal experience and/or research on local practices of sharing and informal distribution. At the Nokia Research Center in Nairobi, for instance, researchers drew on their research on music in Nairobi's slums to develop plans for a system involving person-to-person transfer of music files using Bluetooth.[27]

PHASE 3

In October 2023, I got on a Zoom call with Eric Musyoka, a Kenyan producer with whom I had spent many hours during my fieldwork a decade earlier. Musyoka had just been named as chairperson of a trade association called Recording Industry of Kenya (RIKE), and I was eager to get his new bird's-eye view of the industry. He told me that there had been a major shift since we had last spoken. "The mobile market is still there," he said. "But you see, now, everybody has a smartphone, and, like most of the companies, most of the services that started the revolution through mobile phones, most of them have gone out of business, and we are pretty much left to the big players—the iTunes, the Spotifys, the Boomplays, the Audiomacks."[28]

Available data suggest that the caller ringback tone market in Kenya is still quite a bit larger than the music streaming market in terms of domestic users and overall revenue.[29] These data don't give the full picture, however. Most importantly, they don't capture what is happening with music on social media platforms. They also fail to account for the fact that a portion of the top-earning content on Skiza is not actually music but inspirational speeches and sermons. And they don't reveal the fact that a top-earning artist in East Africa can now bring in around the same revenue from streaming as they can from Skiza—in both cases, between US$7,000 and US$10,000 per month.[30] In any event, for Musyoka and other Kenyan recording industry stakeholders I have spoken with in recent years, Kenya has clearly entered an era in which listeners are discovering and accessing Kenyan music primarily via music streaming services (referred to as DSPs, or digital service providers, by industry professionals) and social media platforms.

The advent of phase 3 of musical platformization was generally foreseen by Kenyan recording industry professionals a decade ago. Multiple interviewees during my extended fieldwork in 2011 and 2012 discussed the inevitability of streaming becoming a primary form of music distribution once smartphone uptake and data costs reached their projected thresholds. Arguably, this point has now arrived. But there is a caveat. While the cost of mobile data in Kenya is, indeed, relatively affordable compared to other African countries and the rest of the world,[31] Mdundo's Martin Nielsen and Boomplay's Martha Huro both noted that data costs had not dropped as precipitously as analysts had predicted a decade ago. This has been a major difficulty for local streaming services, hindering their growth and adoption.[32]

The "big player" streaming services that Musyoka mentioned vary in type. Apple Music (the streaming-oriented successor to iTunes, which Musyoka probably meant to name) and Spotify are leading global platforms—"global" here meaning "not tailored to a specific market or region but rather addressed to a 'universal' consumer."[33] Spotify formally entered the Kenyan market in 2021. In addition to licensing content for local streaming and offering subscriptions priced for local consumers with local payment options, the company set up a local office. It also began working to garner and curate local musics and otherwise respond to regional tastes. Audiomack is a major global streaming service of a different sort—an ad-supported, free platform. Building on its long-standing approach of focusing on specific genres,[34] the company entered Kenya and other African countries through strategic partnerships, including one with the East Africa–focused digital music distributor Ziiki (see below). Finally, Boomplay is unique among the "big players" Musyoka mentioned. Entering Kenya in 2016, shortly after its launch in Nigeria, it is an Africa-focused platform developed by a subsidiary of the Chinese company Transsion, Africa's top smartphone manufacturer. It is neither the first nor the only Africa-focused DSP. However, it is by far the most successful, with a good claim to being "Africa's largest digital music service."[35] Its meteoric rise has been driven, in part, by its rollout as a preinstalled app on Transsion's Tecno devices.

Boomplay and Mdundo occupy a similar space in the Kenyan market but serve different functions. Boomplay is a streaming platform, while Mdundo focuses on downloads. Both platforms employ a freemium subscription model and offer products tailored to the local market, including "DJ mixes." Both also situate the idea of solving market failure at the core of how they represent themselves to the recording industry. Boomplay Kenya's head of content acquisition, Elizabeth Karuru, expressed this plainly to the *Music in Africa* blog:

> "One reason Kenyan music isn't getting around the continent as it should is distribution," Karuru said. "Here at Boomplay we want to solve that problem by formulating new relationships with artists and assuring the industry that we will use our networks and branches across Africa to promote Kenyan music across our platforms."[36]

Additionally, during interviews with me, leaders of both organizations mentioned "piracy," rather than other DSPs, as their "biggest competition."[37]

Where Boomplay departs from Mdundo is in its stance toward engaging with the industry beyond its primary role as a DSP. The reason Karuru was speaking to *Music in Africa* in the first place was to plug a large conference that Boomplay was hosting in Nairobi, which was "aimed at improving the state of music distribution in Kenya's digital space."[38] This was just one of many such industry events that Boomplay has held. Recently, the company has moved into hosting music festivals as well, launching an annual festival called Boomfest in 2024. In carrying out such "industry patronage"[39] activities, Boomplay has essentially taken up the mantle from Safaricom, which was heavily engaged in such activities at the height of phase 2.

Boomplay has been investing in production, too. One Kenyan producer and label owner, Timothy Boikwa, whose career I have been following since 2011,[40] credited Boomplay for allowing his business to thrive during the COVID-19 lockdowns by financing three albums for Kenyan gospel star Mercy Masika, each costing around US$20,000.[41]

At a music industry panel discussion held at the Goethe-Institut Nairobi in 2024, Martha Huro told attendees that Boomplay is "forced to care" about matters of production and marketing because the recording industry is disorganized. As a DSP, she told the audience, Boomplay should really just be "a supermarket. We are not supposed to care. But because we are in a market that is highly

[dis]organized, then we have to care."[42] Huro expounded on this point in an interview with me days earlier, noting, "We've encountered some challenges that are in the market that don't necessarily have to be handled by us, but since we're in the market, it is, it becomes our problem." She went on to lay out a set of talking points about Boomplay's engagements with "policy, education, and finance." By "policy," Huro primarily meant antipiracy advocacy. She explained how she had become vice chair of a group called Partners Against Piracy, which released a report in 2022 asserting that the "Kenyan creative economy is losing 252 million [Kenyan shillings] per day" to piracy. By "education," Huro meant participating in fora like the Goethe-Institut workshop as well as directing support to music education in schools. And by "finance," Huro meant directing capital to the production side of the industry, such as with Boikwa's projects during the pandemic.

The New Content Firms

When Huro appeared at the Goethe-Institut event in Nairobi, she was joined on the panel by two other industry professionals: Agnes Adhiambo Opondo, licensing manager at Mdundo; and Beth Achitsa, artist and label relations manager for Kenya at the Orchard. The Orchard, a subsidiary of Sony Entertainment since 2015, is not a DSP but rather a "digital distributor." While Achitsa's presence on the panel was partly due to her personal relationship with the panel organizer, music publicist and podcaster Anyiko Owoko, it also reflected the growing importance of digital distributors in phase 3 of musical platformization in Kenya.

Digital distributors have two "core roles": uploading music content to digital platforms and distributing royalties to rightsholders.[43] Increasingly, however—as we will see—they also provide guidance and services in the areas of "distribution strategy and marketing."[44] Some are "open platforms" that provide distribution services to any music rightsholder for a fee, while others work with labels or individual artists through negotiated contracts. While music rightsholders can bypass these intermediaries entirely when posting music content on social media platforms, most streaming services—including, as of recently, Boomplay—require that music content be uploaded by a distributor to ensure that the metadata for every song and album, including the essential International Standard Recording Codes, are input properly. Digital distributors have thus found a secure foothold in the Kenyan recording industry.

In recent years, several global digital distributors have entered the East African market for the first time or expanded operations in the region. In 2019, the Orchard made its first local hire on the African continent with Beth Achitsa in Nairobi. Other distributors began staffing regional offices shortly thereafter—including ONErpm, which also set up shop in Nairobi. Many of these companies have followed the lead of Spotify and Boomplay in hiring women in the most visible roles, creating a remarkable shift in which women have increasingly become the face of the music business in Kenya. A full analysis of how regional directors of digital distribution companies and DSPs in Africa have come to be filled by women deserves a study of its own. I can say that women were not absent from key roles in musical platformization in Kenya prior to phase 3; they merely became more visible with its advent. All the women leaders of phase 3 I have interviewed previously worked as music agents, managers, journalists, or A&R professionals during phase 2.

Of the nine digital distributors I am aware of that handle music content from Kenya, only two—Africori and Ziiki Media—were founded in Africa. Both are now affiliated with the multinational Warner Music Group, a key part of Warner's recent strategic expansion into Africa.[45] Ziiki is especially relevant to Kenya, as its focus is East Africa. Founder and CEO Arun Nagar, a Swahili-speaking East African of Indian descent, began his career in digital media distribution as CEO of the African subsidiary of Indian digital content aggregator Spice VAS. In this role, he oversaw the launch of a streaming service called Mziiki in 2014 and then became CEO of Ziiki Media (essentially a rebranding of Spice VAS Africa's content division) in 2019.

Even beyond their obvious role in bringing content to platforms with a global reach, digital distributors have been key to establishing new possibilities for Kenyan popular music to enter global circulation. Phase 3 has introduced some obvious new pathways to global success for Kenyan artists in the form of "spotlights" and "featured playlists" on Apple Music and Spotify. But at the same time, digital distributors have quietly introduced highly targeted strategies. This is exemplified in the story of Tanzanian artist Mavokali's song "Commando," which became popular on TikTok and streaming platforms in 2023. According to Bilha Ngaruiya, Kenya's country manager for ONErpm, the song first became a hit in Portugal and France before catching on in East Africa, thanks to behind-the-scenes efforts by ONErpm. She explained how Martin Price, head of global expansion for ONErpm, had noticed the song trending in Portugal, one of the biggest markets for ONErpm, and asked the Portugal office to use its contacts with DSPs to get the song "playlisted" there. Ngaruiya went on to explain how her office is seeking to replicate this strategy by incorporating small-scale marketing through "influencers."[46]

Reconfiguring the Artist

Digital distributors' engagement with artists in Kenya is not only a matter of collaboration and partnership. These companies are also transforming the role of the artist itself. This is nicely revealed in the story of how Ziiki moved from being a content provider for Skiza to a digital distributor under the umbrella of Warner Music. In addition to amending rights agreements with its artists and bringing on digital distributor Believe as a temporary third party, Ziiki's evolution into a digital distributor involved getting its artists to adopt new social media strategies that would not only generate direct revenue but also enable Ziiki to more effectively market content. To accomplish this, Ziiki used "lucrative advances" to make artists "pay attention to what we had to say." Once "we put the money on the table," recounted Bilha Ngaruiya, who was with Ziiki at the time, the company could walk the artist through such matters as how to situate a catchy segment of a song on TikTok and market it from there. "All of those things took a lot of learning for them," she noted.[47]

In addition to direct work with artists, digital distributors in Kenya work to shape artists' strategies and practices through workshops like the one that took place at the Goethe-Institut Nairobi, which can be clearly recognized as sites of interpellation from a social scientific perspective. The information and advice doled out at these events extends beyond dry descriptions of processes and policies, into what an independent, DIY artist needs to do to succeed. A major focus is *professionalism*. At the Goethe-Institut, Beth Achitsa broached this topic through a discussion of gengetone, a new style of Kenyan hip hop that skyrocketed in popularity around 2019, receiving regional and international attention before quickly declining. Gengetone's sustainability problem, Achitsa claimed, boiled down to the fact that the artists were upstarts who lacked any understanding of how to use contracts or secure proper licenses for samples. This unprofessionalism, she argued, made it difficult to properly market the music and, in some cases, maintain its presence on platforms.

While the Goethe-Institut event involved some discussion about the need for artists to build a team of professionals, the panelists emphasized that artists should control every role. Huro focused particularly on branding, asserting that artists need to develop their "context" through the production of narratives and imagery. At no point did the panelists suggest that getting signed to a label was the right path for an independent artist. What they laid out, instead, was a vision of an independent artist who conforms to the neoliberal ideal of a radically self-sufficient creative entrepreneur.[48] All this advice was surely realistic and helpful for those receiving it. At the same time, it was oriented toward reconfiguring the role of the music artist to benefit the powerful actors in the current phase of musical platformization.

CONCLUSION

In a recent article, Robert Prey and Seonok Lee argue that a "truly *global* understanding of cultural production in an era of online platforms" requires understanding that the process takes different paths in different locales, depending on "political economic and sociocultural contexts of cultural production."[49] They offer a typology of platformization focusing on three dimensions: platform dependence, dominance of "global" platforms, and the degree of platform and recording industry integration. The Kenyan case clearly demonstrates the nuances that may be at play in all these dimensions, particularly the third. On the one hand, the degree of platform and recording industry integration in Kenya is *low*, in that platforms and aggregators have never had a great amount of control over production. But it is also *high*, in the sense that these entities have participated in shaping and reshaping the recording industry over the past decade or more.

What is missing, if anything, from Prey and Lee's typology is the dimension of time. If there is one thing I have tried to capture in the preceding account, it is the tempo and temporality of musical platformization in Kenya. This is because the primary lesson I have learned from following the evolution of Kenya's music recording industry over the past decade is that musical capitalism, like all other forms of capitalism (and capitalism writ large, if we accept the existence of such a thing), is always on the move. It is, to quote Georgina Born, a system in which "new subjects and objects are drawn in, new agencies discovered, new maneuvers adopted, transforming the relations between what is inside and outside, calculable and disavowed."[50]

NOTES

1. David Nieborg and Thomas Poell, "The Platformization of Cultural Production: Theorizing the Contingent Cultural Commodity," *New Media and Society* 20, no. 11 (2018): 4275–92.

2. Christiaan De Beukelaer and Andrew J. Eisenberg, "Mobilising African Music: How Mobile Telecommunications and Technology Firms Are Transforming African Music Sectors," *Journal of African Cultural Studies* 32, no. 2 (2020): 195–211.

3. I began this research under the auspices of the ERC Music, Digitisation, Mediation project, led by Georgina Born (ERC project number 249598).

4. Georgina Born, introduction to *Music, Sound and Space: Transformations of Public and Private Experience*, ed. Born (Cambridge: Cambridge University Press, 2013), 51.

5. Philippe Bouquillion, Christine Ithurbide, and Tristan Mattelart, eds., *Digital Platforms and the Global South* (New York: Routledge, 2023); Zhongwei (Mabu) Li and David Hesmondhalgh, "From P2P to the Cloud: Music, Platformization, and Infrastructural Change in China," *Chinese Journal of Communication* (2024): 1–17; Robert Prey and Seonok Lee, "A Global Approach to Studying Platforms and Cultural Production," in "Global Perspectives on Platforms and Cultural Production," special issue, *International Journal of Cultural Studies* 28, no. 1 (2024): 30–38.

6. See Joost Poort, "21 for 2021: Imagine No Possessions: From Piracy to Streaming," *CREATe blog*, April 14, 2022, www.create.ac.uk/blog/2022/04/14/21-for-2021-imagine-no-possessions-from-piracy-to-streaming/.

7. On reintermediation, see among others Andrew Leyshon, "Time-Space (and Digital) Compression: Software Formats, Musical Networks, and the Reorganisation of the Music Industry," *Environment and Planning A* 33, no. 1 (2001): 49–77.

8. This system began to collapse only recently, defeated not by piracy but by the economic and social disruptions of the COVID-19 pandemic (George Ouma, Facebook Messenger communication, May 28, 2024).

9. Fakii Liwali, recorded interview, July 30, 2012.

10. Richard Njau, recorded interview, January 1, 2012. The backstories of Mwafrika.com, PewaHewa.com, and, to some extent, Kenyanmadness.8k.com are disclosed in Richard Njau's interview with David Kuria on Njau's YouTube channel, Cleaning the Airwaves, "312. Mwafrika.com, East Africa's 1st Gospel Entertainment Website—David Kuria (The Play House)," at 7 min., 41 sec., posted March 9, 2021, https://youtu.be/Xwkvr3h-fcY.

11. Amon Kipyegon Kirui, Mellitus N. Wanyama, and Wilson O. Shitandi, "Disruptive Innovation: Exploring the Impact of Skiza Tunes on the Kenyan Music Industry," *Journal of Music and Creative Arts* 1, no. 1 (2022): 1–10.

12. Andrew J. Eisenberg, "M-Commerce and the (Re)making of the Music Industry in Kenya: Preliminary Notes and Findings" (paper presented at Association for Social Anthropology, New Delhi, April 5, 2012); De Beukelaer and Eisenberg, "Mobilising African Music."

13. Jevans Nyabiage, "Venture Capital Flows to Tech Startups," *Daily Nation*, November 9, 2010, https://nation.africa/kenya/business/technology/venture-capital-flows-to-tech-startups--745038.

14. "Upwardly Mobile: Kenya's Technology Start-Up Scene Is About to Take Off," *The Economist*, August 25, 2012, www.economist.com/business/2012/08/25/upwardly-mobile.

15. Keith Negus, "From Creator to Data: The Post-Record Music Industry and the Digital Conglomerates," *Media, Culture and Society* 41, no. 3 (2019): 367–84. On caller ringback tones defining one's personality, see "Ringtones Promise Big Cash, but Betray Personality Traits," *Business Daily*, May 20, 2009, https://www.businessdailyafrica.com/bd/lifestyle/society/ringtones-promise-big-cash-but-betray-personality-traits-1939914.

16. Negus, "From Creator to Data," 376.

17. De Beukelaer and Eisenberg, "Mobilising African Music," 206.

18. Nanjira Sambuli, recorded interview, May 1, 2012.

19. Josephine Mosongo and Bonface Nyaga, "Behind the Skiza Tunes, Artistes Deal," *Daily Nation*, August 22, 2015, https://nation.africa/kenya/life-and-style/buzz/behind-the-skiza-tunes-artistes-deal-1121938.

20. De Beukelaer and Eisenberg, "Mobilising African Music."

21. De Beukelaer and Eisenberg, "Mobilising African Music," 197.

22. Eisenberg, "M-Commerce and the (Re)making of the Music Industry in Kenya."

23. This information comes from an October 14, 2010, Safaricom press release, archived at https://eastafricabusiness.blogspot.com/2010/10/safaricomhomeboyz-launch-unique-music.html.

24. Russell Southwood, "Content Delivery May See Power Shift from Telcos," *East African*, January 4, 2014.

25. Martin Nielsen, recorded interview, Nairobi, March 23, 2015.

26. Martin Nielsen, recorded interview, Nairobi, March 23, 2015.

27. Moses Sitati, unrecorded interview, October 3, 2011.

28. Eric Musyoka, recorded interview, October 18, 2023.

29. According to Statista, the number of Kenyan subscribers is projected to reach 2.6 million in 2024, up from 1.8 million in 2020. Revenue is projected to reach US$21.63 million in 2024, up from US$12.32 million in 2020. See "Music Streaming—Kenya," Statista, accessed June 5, 2024, www.statista.com/outlook/dmo/digital-media/digital-music/music-streaming/kenya.

30. Timothy Boikwa, WhatsApp conversation, May 30, 2024.

31. According to a 2023 study, Kenya ranks among the twelve African nations where mobile data costs less than US$1 per gigabyte, with a rate of US$0.84. See Anna Fleck, "Worldwide Mobile Data Pricing 2022," Statista, January 19, 2023, www.statista.com/chart/29144/cost-of-mobile-data-in-africa/.

32. Martin Nielsen, recorded interview, April 24, 2024; Martha Huro, recorded interview, May 1, 2024.

33. Prey and Lee, "Global Approach," 32.

34. Benjamin Chertoff, "Audiomack Just Solved Music Streaming," *Business of Business*, March 20, 2018, www.businessofbusiness.com/articles/audiomack-is-quietly-growth-hacking-streaming-music/.

35. Butchie Seroto, "Boomplay Head of Marketing: We're Always Ahead of the Streaming Game," *Music in Africa*, February 7, 2020, www.musicinafrica.net/magazine/boomplay-head-marketing-were-always-ahead-streaming-game.

36. Lucy Ilado, "Boomplay to Host Digital Music Conference in Nairobi," *Music in Africa*, May 17, 2019, www.musicinafrica.net/magazine/boomplay-host-digital-music-conference-nairobi.

37. Martin Nielsen, recorded interview, April 24, 2024; Martha Huro, recorded interview, May 1, 2024.

38. Ilado, "Boomplay to Host Digital Music Conference in Nairobi."

39. De Beukelaer and Eisenberg, "Mobilising African Music."

40. See Andrew J. Eisenberg, "Soundtracks in the Silicon Savannah: Digital Production, Aesthetic Entrepreneurship and the New Recording Industry in Nairobi, Kenya," in *Music and Digital Media: A Planetary Anthropology*, ed. Georgina Born (London: UCL Press, 2022), 77–82.

41. Andrew J. Eisenberg et al., "Locking Down the Beat: Musical Livelihoods in Nairobi during the Covid Pandemic (Audio Documentary)," YouTube, 50 min., 45 sec., posted November 21, 2022, by Music and Sound Cultures, https://youtu.be/t9PmvJ8HoYI&t=899s.

42. Goethe-Institut Nairobi, "Streaming Revenues and Music Distribution," Facebook video, 2 hr., 52 min., 51 sec., May 10, 2024, www.facebook.com/GoetheInstitutNairobi/videos/3256889231273682.

43. "The Mechanics of Distribution: How It Works, Types of Music Distribution Companies and 35 Top Distributors" *Soundcharts*, December 31, 2023, https://soundcharts.com/blog/music-distribution.

44. "Mechanics of Music Distribution."

45. Murray Stassen, "Warner Acquires Africa-Based Music Distributor Africori," *Music Business Worldwide*, January 12, 2022, www.musicbusinessworldwide.com/warner-acquires-africa-based-music-distributor-africori1/.

46. Bilha Ngaruiya, recorded interview, April 24, 2024.

47. Bilha Ngaruiya, recorded interview, April 24, 2024.

48. Andrea Moore, "Neoliberalism and the Musical Entrepreneur," *Journal of the Society for American Music* 10, no. 1 (2016): 33–53.

49. Prey and Lee, "Global Approach," 36.

50. Georgina Born, "Future-Making: Corporate Performativity and the Temporal Politics of Markets," in *Cultural Politics in a Global Age: Uncertainty, Solidarity and Innovation*, ed. David Held and Henrietta L. Moore (London: Oneworld, 2007), 295.

3

Music Streaming, Platform Labor, and Intermediaries

Emília Barna

INTRODUCTION

The entry of IT corporations into the music economy through online music and video streaming platforms has arguably been the most forceful development in the global music industries of the last two decades. Countless scholarly accounts have explored how these platforms have shaped music consumption[1] and musicians' labor, including income and working conditions.[2] Elements of a democratization narrative have partly appeared in such accounts, particularly the idea that streaming platforms have provided independent or aspiring, "semiprofessional" musicians with the means to distribute their music to an international public.[3] At the same time, studies have also criticized the overall low income from streaming, the impact of platform metrics on musicians' work and self-valuation,[4] and the various ways in which platforms reinforce existing inequalities in the music industries.[5] Platforms have been described as the primary new intermediaries between musicians and consumers, actively shaping this relation. Yet less attention has been paid to actors mediating between streaming platforms and musicians, even though they influence musicians' working conditions, income, and opportunities. Some of these intermediaries are "new," such as digital distributors (sometimes called aggregators), and some are "old," such as collecting societies. Some are for-profit, while others—again, collecting societies and trade organizations—are nonprofit. Some are transnational—global or regional—while others are national-level. They are all, however, embedded into local music industry histories and (infra)structures, as well as global-level power relations among industry players.

This chapter explores the ways in which new intermediaries of the digital music streaming ecosystem, both local and transnational, shape the work of musicians locally. By focusing not only on streaming platforms but also on distributors and collecting societies as intermediaries, I shed light on actors that receive less emphasis in studies of the digitalization of music. Nevertheless, these actors crucially shape musicians' access to streaming platforms in (semi)peripheral countries where music markets are "small" and the presence of major record labels has been less significant. Through a theoretical perspective drawing on the political economy of cultural production, labor process theory, critical accounts of cultural labor and its platformization, and a critical understanding of global power relations with the help of world-systems analysis, I explore the ways in which intermediaries channel local labor into streaming platform-based music production.

I focus on Hungary, an Eastern European country considered semiperipheral from a world-systems perspective, occupying "an intermediate position in the core-periphery structure of the capitalist world-economy."[6] The Hungarian music industries are strongly shaped by global trends of capital concentration in the cultural and IT industries, with musicians and other music industry actors positioned between local demands—such as a demand for Hungarian-language popular music addressing a local audience—and global competition. The structural transformation of the Hungarian music industries, particularly the recording industry (the dominant segment in terms of economic value at the time), after the 1989–90 regime change was defined by integration into a global market dominated by major record labels, as well as the rise of various local players such as Hungarian record labels, management, media production (popular music radio stations and music television), and live music enterprises. The rise of digital streaming has similarly facilitated the development of a local infrastructure, which serves as a ground where power relations and struggles between the capital accumulation of corporations with a global reach—digital platform companies—and local labor may be observed. In the following, first, I look at how local players are embedded in the global industry; second, I explore how distributors, as intermediaries, channel local musicians' labor into the global economy of music; and third, I reflect on practices and strategies of local musicians facing control and governance by streaming platforms and music industry workers in the form of subordination, alignment, relative autonomy, and resistance.

The analysis relies on eleven semistructured interviews with employees of digital distributors (4, representing three companies), record label employees (2), musicians (3, one of whom is also a record label employee), and representatives of the three Hungarian collecting societies: EJI (Bureau for the Protection of Performers' Rights), Artisjus Hungarian Bureau for the Protection of Authors' Rights, and MAHASZ (Association of Hungarian Record Companies, which functions as both a collecting society and a trade organization for record labels). The transcribed interviews were coded and analyzed, along with various local and global industry and trade documents (music industry reports), accounts from the Hungarian music industry press, and presentations at an industry event organized by the distributor Believe Digital in Budapest. With the help of this data, I aim to contribute to a global-perspective understanding of new conditions for music creators, focusing on labor in music and the role of intermediaries.[7]

THEORETICAL APPROACH

In the recording industry, content-based digital platforms, including streaming services, signify a process of "reintermediation."[8] These platforms, run by IT corporations such as Google (YouTube) or Apple (iTunes, Apple Music), along with the music streaming platform Spotify, increasingly mediate the relationship between creators—primarily musicians—and their audiences. As Hyojung Sun argues, "Quite distinct from the widely perceived prediction that the recording industry would experience a radical change through disintermediation, the digital music industry is more reintermediated than ever before."[9] The idea of reintermediation responds to initial hopes of "disintermediation" expressed in more techno-optimistic accounts of internet-based music production and consumption regarding the possibilities offered by digital and online technology to circumvent traditional gatekeepers, particularly record labels and mass media such as television, radio, and the press. Indeed, a significant segment of musicians worldwide are capable of building "DIY" careers[10] without a record label contract; however, they rely on new intermediaries for the distribution and sale of their music. Although "older" gatekeepers have not disappeared, their functions and power have shifted in accordance with a new industry structure where actors profiting from music now include digital platform companies.

Since digital content-based platforms significantly and increasingly mediate recorded music on a global scale, we need to consider how they influence work in the music industry. Analyses aimed at the platformization of cultural and media labor have mainly highlighted the organizing power of algorithms and the lack of transparency in terms of remuneration and other aspects such as visibility,[11] the power and influence of ranking and rating systems,[12] the ways in which they enhance entrepreneurialism[13] as well as competition among individual workers,[14] and governance[15] from the side of the platform. The theoretical tradition of the labor process and its application to work in various contexts, from factories to the platform-based "gig economy," helps to highlight the relationship between technology and labor in terms of control, exploitation, and alienation on the one hand, and the conditions of worker autonomy, creativity, and resistance on the other. Based on Karl Marx's *Capital: Volume One*,[16] where he outlined the relationship between the creative power of human labor and the capitalist mode of production, Andrew Friedman[17] described the organization of industrial labor in the context of the factory through managerial strategies taking the form of "direct control" and "responsible autonomy."[18] The labor process perspective has since been applied to understanding work in the digitalized cultural and creative industries. Alessandro Gandini,[19] for instance, explores emotional labor and control in the digital platform-based gig economy, paying particular attention to ranking and rating systems that serve purposes of managerialization and monitoring of workers. Ellis Jones,[20] moreover, explores DIY musicians' use of social media platforms, based on Harry Braverman's[21] work. The "scientific management" of Taylorism parallels, according to Jones, the "nudging" logic of platforms, where it is "the monopolistic rulers of platform capitalism who are best placed to nudge."[22] Critical reflections on platform-based entrepreneurship tend to highlight how workers come to embrace neoliberal, individualized strategies in their work and careers. These accounts reveal that entrepreneurialism as a discourse,[23] and entrepreneurial subjectivities themselves,[24] tend to celebrate and embrace platforms as a technological toolkit—a means of production—and a set of opportunities, without considering their exploitative, individualizing, and potentially alienating logic.

Addressing the ways in which new intermediaries mediating between platforms and creators shape the latter's labor is largely missing from these accounts. An important exception is Michael Siciliano,[25] who provides a compelling theorization of creative labor from a labor-process perspective based on two cases: workers in a recording studio; and YouTube content creators, along with intermediary workers, at a so-called multichannel network (MCN)[26] assisting and managing such creators' work. Patryk Galuszka[27] provides a detailed account of the functions and operation of digital distributors from an Eastern European perspective, emphasizing that aggregators bundle digital rights—copyright and performers' rights—to deliver them to digital music stores[28] and that they "help resolve . . . the bargaining asymmetry that exists between large digital music stores and small independent music labels or individual artists."[29]

One aspect missing from both of the above, however, is an examination of the global organization of the digitalized media and music industries. The platformization of music production, distribution, and consumption has dominantly been interpreted from perspectives rooted in the global core. At the same time, media platforms are global in the sense that "content" is produced and consumed worldwide, including in peripheral and semiperipheral countries, with labor divided along global power relations.[30] Some early, prestreaming accounts of the digitalization of music from a geographical perspective are highly instructive. In 2006, Gustavo Azenha[31] convincingly argued that digitalization would likely reinforce the already existing trend of diversification in the recording industry that has been accompanied—perhaps counterintuitively—by the concentration, including geographically, of distribution, licensing, and marketing.[32] He further contended that digital technologies would ultimately reinforce existing socioeconomic power relations through digital distribution.[33] Andrew Leyshon[34] also showed that the digitalized music network had remained predominantly Western-based, at least from the point of view of capital.

The global organization of platformized cultural and media labor also means that local labor is channeled from different positions into its global value chains. Digital platforms have been shown to create a renewed exposure of creative workers to the logic of capital concentration, ultimately reinforcing geographical inequalities.[35] In this chapter, I argue that the allocation of creators'—including musicians'—income depends on access to both local or international distributors and other intermediaries, such as national collecting societies. The existence of specific deals, or the lack thereof, between such intermediaries and platforms crucially affects working conditions locally.

Among Western-based studies and a growing number of accounts of music streaming in large semiperipheral markets such as China[36] or India,[37] perhaps most overlooked are smaller semiperipheral states, such as those of the Eastern European region. The semiperiphery is a diverse category, comprising countries combining "a near even mix of core-like and peripheral" production processes[38] and mediating between core and periphery.[39] Stemming from this intermediary position, global transformations tend to affect semiperipheral states strongly, and they may also occupy strategic positions in technological innovation and development.[40] I will argue through the example of Hungary that the Eastern European semiperiphery has indeed played specific strategic roles in the development of the platformization of recorded music, particularly as a source of knowledge capital and as a testing ground.

GLOBAL PLATFORMS AND LOCAL INTERMEDIARIES ON THE EASTERN EUROPEAN SEMIPERIPHERY

Although internet access, especially broadband, spread at a slower pace in the region compared to the United States or Western Europe, by the early to mid-2000s, internet-based technologies of music production, distribution, and consumption had also become widespread in Hungary. The global recording industry crisis in the first half of the 2000s led to major labels closing many local offices around the world. As part of this, all majors shut down their offices in Hungary except Universal (which remained until 2023). The decline in physical sales was drastic in Hungary, as it was globally, and the growth of income from digital music—which already began globally in 2004[41]—occurred later and at a slower rate in Hungary. According to data published by MAHASZ, 2013 marked the low point in combined digital and physical sales.[42] YouTube launched monetization in Hungary in 2012, aided by its so-called Content ID system, and it remains the strongest platform up to the present day. (Content ID had been developed by 2007, meaning that "YouTube was relatively fast to launch it [in this country].")[43] The French music streaming company Deezer entered the Hungarian market in the same year, and Spotify in 2013. Together with Apple Music, these constitute the biggest streaming platforms for musicians and music consumers locally. The appearance of digital distributors also followed monetization, including the Hungarian WM Music Digital (WMMD) and a small number of international companies opening local offices or regional ones with local representation, such as Believe Digital (based in France) in 2013 and the state51 Conspiracy (based in the United Kingdom) in 2019. Many Hungarian musicians partner with "DIY distributors," transnational companies without any local embeddedness (i.e., no local office or locally available contact): TuneCore, CD Baby, DistroKid, RouteNote, Ditto, and ReverbNation were mentioned in the interviews. The former types of distributors—WMMD, Believe, or the state51 Conspiracy, which I will refer to as locally embedded companies—strongly distinguish themselves from "DIYs" according to the types of deals and services they offer: they usually take a greater percentage (depending on the "package") but offer personal contact and various personalized services beyond monetizing music on streaming platforms (one employee, for instance, mentioned getting a musicians' track on a local radio station).

Regardless of the recent growth of the digital sector, live music still dominates Hungarian musicians' income.[44] Moreover, as the interviews also confirmed, streaming numbers significantly rely on live music: playing at major music festivals and venues is crucial for artists in building an audience. The COVID-19 period nevertheless brought a turn: by 2023, streaming subscriptions had significantly increased, moving away from the predominance of free platforms, particularly YouTube. (At an industry event organized by Believe Digital, the company reported an increase in streaming subscribers from 5.9 percent of the population in 2019 to 18.2 percent in 2023.)[45] Moreover, the first years of the 2020s also brought a turn in the international versus local character of streamed music: when MAHASZ began releasing digital sales charts in 2014, Hungarian artists were scarcely represented. Today, however, international music typically makes up less than 10 percent of the digital Top 40.[46] This shift corresponds to a global trend of increased consumption of local music on streaming platforms.[47] The interviewed music industry workers often described the shift in Hungarian music streaming as a generational change, marked by musicians adopting a more proactive approach and achieving quicker success:

> COVID brought a change in all respects from the perspective of artists, a whole generation, and a generation with completely different attitudes. The fact that the evolution of a band is two to three years—it's not only that they miss [the stage of] crawling on all fours: they're practically born and running in two years.[48]

It is important to emphasize, however, that this period of growth is relative, and income from streaming remains highly unequal. Only a handful of successful musicians generate considerable earnings, and even for them, live music is often the primary source of income. As an example, a Hungarian musician frequently cited as successful in streaming confirmed that their income came "overwhelmingly [from] live music."[49] "The majority of musicians," as a representative of EJI put it, "receive pennies"[50] from streaming.

The opportunities of musicians based in Hungary partly depend on the positions occupied by distributors in relation to platform companies. Platforms allocate "preferred partner" statuses to some distributors, distinguishing them from "DIY distributors." As an employee of one of the main distributors—an international company with a local office—explained, DIY distributors do not filter content and often allow fraudulent recordings, which are later taken down (e.g., after Content ID analysis on YouTube). By contrast, companies such as the one in question perform prefiltering work, thereby earning their highlighted status. As a result of this status, "one stream is calculated based on the biggest revenue per mille."[51] Their position, however, does not reach that of major labels and therefore does not match the income that majors and affiliated distributors such as Universal's Virgin—also recently reappearing in Hungary—may offer artists. ("Whoever is on the market, one thing is certain: that Universal has the best share with Spotify; this is unquestionable.")[52] Existing inquiries into income inequalities—or the generally low income—from streaming among musicians rightly point to the role of record labels and the deals they hold with musicians;[53] yet the above indicates that distributors' position on the global streaming market also matters.

Even more significantly, earnings are also shaped by factors associated with local collecting societies, which are relatively small organizations with limited resources. These factors include the specific deals—or lack thereof—between platforms and collecting societies; access to infrastructure for managing the vast quantity of data received from streaming platforms, based on which they are supposed to distribute royalties; and collecting societies' own policies of restructuring income, which they may use to offset, to an extent, the difficulties of data management. As the representative of EJI argued, streaming platform companies are not prepared to deal with performers' collecting societies (as opposed to societies representing author rights).[54] Uniquely, EJI sued Deezer in 2014 for infringing performers' rights by failing to pay performance royalties, winning the lawsuit in 2018.[55] This was followed by another lawsuit against Spotify, which ended in a mutual agreement in 2021. Only after these legal procedures did the two companies begin paying the relevant royalties to the collecting society, which represents nearly all Hungarian professional performing musicians, while income from the remaining platforms remains nonexistent. The difficulties nevertheless continued, as managing the data provided by platforms has posed a significant problem: "The data transferred by a single major streaming service provider to EJI in a month is an order of magnitude greater than the total amount of data previously processed by EJI in a year," as the representative illustrated. The same problem was only solved by Artisjus in 2023 by partnering with a larger French company, SACEM, which helps manage Hungarian songwriters' rights on streaming platforms.

DIGITAL DISTRIBUTORS CHANNELING LOCAL LABOR

Shuwen Qu, David Hesmondhalgh, and Jian Xiao[56] demonstrate how Chinese music streaming platforms channel the labor of self-releasing musicians, who constitute a "reservoir" or "proto-market."[57] In Hungary, self-releasing artists, although often starting out with a so-called DIY distributor, are also actively sought by locally embedded distributors. These distributors partly perform A&R roles, as one of the interviewed employees exemplified, describing the recruitment part of their job thus:

> One of my great findings was a TikTok-trend–like recording popping up in a random Instagram video, which could rather be categorized as a meme, but . . . you could already see at the beginning that this would turn into an enormous trend on the domestic market. This was the song of a young Roma artist, and I quickly did a search on them, and there was no channel, nothing [monetized] anywhere. So I quickly wrote to them. They immediately said, "Wow, this is a wonderful opportunity," and we signed the contract.[58]

Beyond recruitment, distributors proved to be important agents of shaping the labor process and musicians' attitudes, communicating expectations and rewarding certain types of conduct. All the interviewed distributor employees described their most important expectation from a musician partner: exhibiting conscious planning and a strategic attitude. "For us, the ideal partner is one with long-term ideas. To exhibit this 'long-term thinking' [phrase in English in original]. If they don't have that, then unfortunately it's a lost cause most of the time."[59] (The English-language formulation of "long-term thinking" may be an indicator that it comes from an international industry discourse.) WMMD, one of the Hungarian distributors, rewards "good partners" as part of prizes they hand out every year (the rest of the prizes are based on success in terms of statistics, thus more "objective" data):

> The Partner of the Year prize . . . we give to the partner that in the given year, [demonstrates] all of the things I've said before—planning ahead . . . [it is somebody] with whom we feel we can work together the most effectively, who considers our advice and the opportunities, and so on. We've managed to achieve real success together.[60]

As a contrasting negative example, distributor employees cited expectations by musicians that a song just completed should be released immediately. One employee observed that they saw "some really slow improvement" in this respect on the part of musicians, yet they were "dissatisfied in this respect because it is not at a pace that would be ideal, and many opportunities are being missed."[61]

The tendency of the "development" of musicians' mindset in a positive direction is therefore linked to becoming attuned to a specific labor process afforded by streaming platforms, wherein organizing work around a regular, reliable, and consistent output—following an almost mechanical schedule—is key. Moreover, each track needs to be "pitched" carefully—not only to provide sufficient information for potential playlist curators but also as a performance of professionalism and reliability through precision and attention to detail:

> The other important area is Spotify for Artists: the lack [of a profile], or its unsatisfactory appearance. Spotify for Artists is typically the interface that, if the artist wants to get on a playlist—which is what everyone wants—has to be preceded by a pitching process. And unbelievable as this may sound, at Spotify, they do check artists: they will check whether they have a profile, whether it was the artist themselves that pitched the song, what the profile looks like, [and whether they] have . . . at least set a picture.[62]

The employee cited the example of a musician for whom none of their first four or five singles managed to end up on playlists. Eventually one did, which the employee attributed to the displaying of consistency—which, in their view, earned trust from the platform. In this sense, the distributor—through communicating such expectations (which were echoed by the interviewed musicians)—and rewarding musicians adhering to them, performs a form of disciplining musicians' labor on behalf of platforms.

Second, musicians' attitudes were also compared to what distributor employees described as dominant attitudes in the "West"—appearing as a mythical, moral center:

> And this is a great big contrast with, for instance, the British market or the American market . . . , where independent artists, even the smallest ones, are way more conscious in their attitude. . . . As a minimum, they are present on all social platforms, they have proper profile pictures, they claim their Spotify for Artists profile, they have a press photo kit [and] a bio that can be used any time. This, unfortunately, cannot be said about most Hungarian artists, and there is a great deficit in planning as well.[63]

This comparison, unfavorable to Hungarian artists, can be interpreted within a framework of what József Böröcz[64] terms moral geopolitics, where professionalization in the music industries is embedded in a discourse of "catching up" with the Western core. It corresponds to the geography of the flow of industry knowledge: industry professionals—some equipped with training or music industry degrees from the United Kingdom or other Western European countries—regularly attend Western European and US industry events to keep their knowledge up-to-date, and then attempt to transform this knowledge, along with the dominant industry discourse from the global core, to partners. Increasingly, for professionals working for distributors positioned close to platforms through the "preferred partner" status, platform companies directly serve as a source of music industry knowledge relating to services and the operation of platforms: "Basically, we have an internal bulletin system, where, because we have this preferred partner status and we have a system of trust with the biggest stores, the head office will tell us in advance about the upcoming changes—internal audits."[65]

In addition to personal communication with partners, distributors' expectations—or, rather, platforms' expectations filtered through locally embedded distributors—were actively communicated in the form of education: for instance, events such as webinars—at times featuring representatives of Spotify and YouTube directly—and (social) media content. For instance, WMMD shares the latest changes regarding platforms via its TikTok and Instagram accounts. This educational work is not only aimed at musicians—although it is reaching them with limited effect, as distributors themselves admit—but also professionals in the Hungarian music industries more broadly: there are frequent collaborations between Believe and Artisjus, or WMMD and MAHASZ. Spreading knowledge and expectations regarding the labor process, mostly originating from music industry actors of the global core and directly from platforms, plays a vital part in shaping a receptive local space that enables its use as simultaneously a resource of specific knowledge and a testing ground for platforms: the employee of Believe, for instance, explained how Spotify ran "the entire beta test" for algorithmic placement with Believe's catalog.[66]

CONTROL AND AGENCY BETWEEN PLATFORMS AND CREATIVE WORKERS

In the final section, I explore positions and strategies that creative workers in the local music industries, including intermediaries and musicians, display in the face of control from streaming platforms. These positions and strategies are responses to the dilemma of art or creativity versus commerce, a definitive element of labor in the cultural industries,[67] but within the specific context of streaming. I distinguish between *subordination* as being in a vulnerable position, without power to question the system (a position with little agency); *alignment* as benefiting from the system, although still from a subordinate position, without questioning it (a position with agency); *relative autonomy* as circumventing control to an extent, but without questioning the system (a position with agency); or *resistance* as circumventing control by questioning the system (a position with agency).

Streaming platforms' expectations regarding "content," produced by musicians and mediated by distributors, affect the creative labor process and the form or aesthetic of the creative product. One problematic point named in some of the interviews was the issue of profile pictures and album art: platforms, according to a distributor employee, "can have a say in what an artist photo [should look like]. . . . They force this sterile background bullshit on everyone, but at the same time, [they] do not contribute a penny to photography."[68] According to the professional, at least one record label from the region (not Hungary) adapted to this expectation by having "put up a whiteboard, and once the recording is completed, they take a photo of their artists according to the guidelines." This indicates that players feel forced to comply from a subordinate position, yet the interviewed professional also exercised relative autonomy in the face of this pressure by not complying in the case of an artist whose image and art, they felt, would have been seriously compromised by producing the "sterile" artwork expected by Apple Music. Images cut from the artist's video would not have conformed to the guidelines due to an emphasis on violence (the artist in question was a rapper, and the video featured tanks and machetes). Instead of requesting alternative photos, the employee of the distributor decided not to pitch it for a playlist but to look for alternative channels of monetization: "This is the artist; don't pretend they're a Milky Chance or Ed Sheeran." Controlling "content" on such a basis can be compared to YouTube demonetizing certain channels or removing videos not conforming to the platform's guidelines around violence or sexual content, which Caplan and Gillespie have interpreted as "private governance by platforms."[69]

The above indicates that employees of digital distributors may exercise some agency in assisting the preservation of relative creative autonomy for musicians. Another area of practicing relative autonomy in relation to platforms is the public sharing of strategies and detailed "analytics"—data on how particular artists performed through particular distributing strategies, such as algorithmic placing—by at least one of the distributors, either at various Hungarian industry events or as part of guest lectures in music industry education. "If transparency is our basic principle, then we simply must speak about things like this, so whenever I'm at conferences, . . . I tell confidential data to the people that are present," they explained.[70] This practice is aimed at educating the broader local industry, not only the distributors' own partners, and involves, at least to an extent, some resistance to the logic of competition. Overall, it nevertheless contributes to establishing more effective ways of channeling artists into the distribution and streaming ecosystem. In another case, however, a distributor aligned itself with the Western industry—positioning itself as "catching up"—while opposing the protection of artists' interests in the previously mentioned collective struggle for performing rights royalty income, essentially taking the side of capital against labor:

> So [EJI] had sued [Spotify and Deezer], while these services are following a perfectly benevolent conduct and process. And [EJI is] imposing a kind of accountability on them that is not a standard music industry process. And it is not [according to the] standard business model of these services, and, in my opinion, that is one of the reasons why Deezer has left Hungary,[71] stopped offering free [subscriptions], and completely lost focus, which I don't think benefited Hungarian artists in the short or long term.[72]

The employee added, for emphasis, that in Germany, YouTube had not monetized music content for seven years due to a similar conflict between the local collecting society and the platform's business model. They emphasized that Hungary is a "statistical error"—meaning invisible and insignificant—in relation to the German market, thus highlighting Hungarian actors' relative lack of power and suggesting a strategy of subordination.

The abovementioned guidelines regarding artwork are direct expectations from platforms, yet platforms may also exert control over the creative process and product through the operation of algorithms. With regard to tailoring composition according to what is likely to succeed on streaming platforms, the positions of the interviewed musicians differed. A songwriter and producer said that he had begun writing consciously with the specific logic of streaming platforms in mind about two years previously. He also observed, however, that "this is a bit frustrating as a musician and composer, but you have to try to get it out of your head, [otherwise] you don't make any progress."[73] In contrast, another musician—the member of a band—insisted on maintaining the band's creative autonomy in songwriting decisions: "No, never, we never considered [streaming platforms] at all, to any extent. And if you look at the entirety of the career path of the band, it might seem that we keep shooting ourselves in the foot a bit every time."[74] The two different attitudes may be explained by the musicians' different roles and strategies of making a living from music: as a producer, the first musician worked with many singers or rappers in many different genres, emphasizing his own flexibility and diverse skill set—"R & B music, electro, hip hop, pop, *mulatós*[75]—anything you can imagine." He also made it clear that the same level of diversity and heteronomy would not have been necessary had he pursued a career as a performing DJ-producer. Yet, without live music income, he had to prioritize flexibility and the demands of the algorithm. The band, in contrast, strongly relied on live music income and a highly loyal Hungarian following, along with a sizable international audience, which they had cemented by playing gigs. A third musician stood in between, confirming that they were aware of the kind of music needed to succeed on specific playlists. However, so far, they had not purposefully done this, though they did not rule out the possibility: "If Lofi Girl,[76] for instance, asked me to do an album in the kind of light style that they like to release, I'd be happy to do one like that."[77] This attitude displays a striving for autonomy together with alignment. The space for choice in the case of this artist is provided, on the one hand, by several economic "legs" to stand on—though all related to music—as well as an established network, including international connections, partly through the niche profile record label for which they worked.

CONCLUSION

A focus on the Eastern European semiperiphery reveals the multifaceted and complex role of intermediaries occupying the space between streaming platforms and both musicians and small record labels. In regions outside the global core and the center of the music and IT industries, musicians' income and opportunities depend primarily on two factors: first, platforms' geographical policies, in particular local representation in the form of offices or local playlists; and second, the particular deals between platforms and distributors, including distributors' position in the global market and deals between platforms and collecting societies. Unequal power relations between global platform companies and local or locally embedded players—for instance, the unequal resources for handling streaming data—thoroughly determine the work of intermediaries, musicians, and other local music industry workers. Nevertheless, EJI's example showed that local players—particularly nonprofit collective organizations serving to represent workers' interests—may exercise agency and fight for better working conditions for musicians by protesting existing platform policies.

The case of music streaming in Hungary manifests a characteristic semiperipheral hybridity in the sense that local partners provide new markets to develop for corporate intermediaries such as digital distributors, who capitalize on local creative labor and various forms of specialized professional knowledge, while also channeling local labor toward the global core and actively shaping the labor process. Through personal communication with partners and educational efforts aimed at professionals of the local music industries, distributors perform work that is functionally being outsourced to them by platforms, who tend to remain distant from the (semi)peripheries. In terms of the desired labor process, consistent output by musicians is critical. Musicians—especially those relatively successful on streaming platforms—have partly internalized these expectations in their work. However, they often remain "reluctant entrepreneurs"[78] and continue to engage in practices and strategies of resistance. The degree of their creative autonomy depends on their income structure and background—the "legs" they must stand on while attempting to build careers in music. Strategies regarding streaming need to be complemented, in most cases, by an active live music presence as well as networking—both locally, which comes with its own inequalities, such as gender relations or dependence on government policies and political connections,[79] and internationally. Examples emerging from the research show that the latter may be successful in particular niche genres, such as lo-fi, but it requires consistent and deliberate effort.

NOTES

1. For a comprehensive literature review, see David Hesmondhalgh, Raquel Campos Valverde, D. Bondy Valdovinos Kaye, and Zhongwei (Mabu) Li, "The Impact of Algorithmically Driven Recommendation Systems on Music Consumption and Production: A Literature Review," UK Centre for Data Ethics and Innovation Reports, 2023, https://ssrn.com/abstract=4365916.

2. David Hesmondhalgh, "Is Music Streaming Bad for Musicians? Problems of Evidence and Argument," *New Media and Society* 23, no. 12 (2021): 3593–3615.

3. See Hesmondhalgh's chapter in this volume (chapter 1).

4. See for example Robert Prey, "Performing Numbers: Musicians and Their Metrics," in *The Performance Complex: Competition and Competitions in Social Life*, ed. David Stark (Oxford: Oxford University Press, 2020), 241–58.

5. See for example Tamas Tofalvy and Júlia Koltai, "Splendid Isolation: The Reproduction of Music Industry Inequalities in Spotify's Recommendation System," *New Media and Society* 25, no. 7 (2023): 1580–1604.

6. Giovanni Arrighi, "The Developmentalist Illusion: A Reconceptualization of the Semiperiphery," in *Semiperipheral States in the World-Economy*, ed. William G. Martin (Westport, CT: Greenwood Press, 1990), 11.

7. In the chapter, I quote from the interviewed professionals anonymously, with distributor companies also mostly anonymized. Where it was important to name the company, I omitted the interview date to disconnect those excerpts from other quotes. All interviews were conducted in Hungarian; the quotations are my translations.

8. See for example Patryk Galuszka, "Music Aggregators and Intermediation of the Digital Music Market," *International Journal of Communication* 9 (2015): 254–73.

9. Hyojung Sun, *Digital Revolution Tamed: The Case of the Recording Industry* (Cham, Switzerland: Palgrave Macmillan, 2019), 237.

10. DIY here is understood not in an underground sense, but as a practice that has been "mainstreamed" in the age of digital media; see Ellis Jones, *DIY Music and the Politics of Social Media* (New York: Bloomsbury Academic, 2021).

11. See for example Brooke Erin Duffy, "Algorithmic Precarity in Cultural Work," *Communication and the Public* 5, no. 3–4 (2020): 103–7.

12. See for example Alison Hearn, "Structuring Feeling: Web 2.0, Online Ranking and Rating, and the Digital 'Reputation' Economy," *Ephemera* 10 (2010): 421–38.

13. See for example Brooke Erin Duffy and Emily Hund, "'Having It All' on Social Media: Entrepreneurial Femininity and Self-Branding among Fashion Bloggers," *Social Media and Society* 1, no. 2 (2015), https://doi.org/10.1177/2056305115604337; Jo Haynes and Lee Marshall, "Reluctant Entrepreneurs: Musicians and Entrepreneurship in the 'New' Music Industry," *British Journal of Sociology* 69, no. 2 (2018): 459–82; Jia Guo, "The Postfeminist Entrepreneurial Self and the Platformisation of Labour: A Case Study of Yesheng Female Lifestyle Bloggers on Xiaohongshu," *Global Media and China* 7, no. 3 (2022): 303–18.

14. Nancy Baym et al., "Making Sense of Metrics in the Music Industries," *International Journal of Communication* 15 (2021): 3418–41; Prey, "Performing Numbers."

15. See for example Robyn Caplan and Tarleton Gillespie, "Tiered Governance and Demonetization: The Shifting Terms of Labor and Compensation in the Platform Economy," *Social Media and Society* 6, no. 2 (2020), https://doi.org/10.1177/2056305120936636.

16. Karl Marx, *Capital: Volume One*, trans. Ben Fowkes (1867; repr., London: Penguin, 1976).

17. Andrew L. Friedman, *Industry and Labour: Class Struggle at Work and Monopoly Capitalism* (London: Macmillan, 1977).

18. On Taylorism, see also Harry Braverman, *Labor and Monopoly Capital* (New York: Monthly Review Press, 1974). On "manufacturing consent" among workers in the factory, see Michael Burawoy, *Manufacturing Consent: Changes in the Labor Process under Monopoly Capitalism* (Chicago: University of Chicago Press, 1982).

19. Alessandro Gandini, "Labour Process Theory and the Gig Economy," *Human Relations* 72, no. 6 (2019): 1039–56.

20. Jones, *DIY Music.*

21. Braverman, *Labor.*

22. Jones, *DIY Music*, 131.

23. Guo, "Postfeminist Entrepreneurial Self."

24. Duffy and Hund, "'Having It All.'"

25. Michael L. Siciliano, *Creative Control: The Ambivalence of Work in the Cultural Industries* (New York: Columbia University Press, 2021).

26. Siciliano, *Creative Control*, 132.

27. Galuszka, "Music Aggregators."

28. Galuszka, "Music Aggregators," 262.

29. Galuszka, "Music Aggregators," 263.

30. See for example Christian Fuchs, *Digital Labour and Karl Marx* (New York: Routledge, 2014).

31. Gustavo Azenha, "The Internet and the Decentralisation of the Popular Music Industry: Critical Reflections on Technology, Concentration and Diversification," *Radical Musicology* 1 (2006), www.radical-musicology.org.uk/2006/Azenha.htm.

32. Azenha, "Internet," para. 79.

33. Azenha, "Internet," para. 5, 26, 122.

34. Andrew Leyshon, "Time-Space (and Digital) Compression: Software Formats, Musical Networks, and the Reorganisation of the Music Industry," *Environment and Planning A* 33, no. 1 (2001): 49–77.

35. Tofalvy and Koltai, "Splendid Isolation."

36. See for example Shuwen Qu, David Hesmondhalgh, and Jian Xiao, "Music Streaming Platforms and Self-Releasing Musicians: The Case of China," *Information, Communication and Society* 26, no. 4 (2023): 699–715.

37. Anaar Desai-Stephens, "The Infrastructure of Engagement: Musical Aesthetics and the Rise of YouTube in India," *Twentieth-Century Music* 19, no. 3 (2022): 444–71.

38. Immanuel Wallerstein, *World-Systems Analysis: An Introduction* (Durham, NC: Duke University Press, 2004), 28–29.

39. Christopher Chase-Dunn, "Comparing World Systems: Toward a Theory of Semiperipheral Development," *Comparative Civilizations Review* 19 (1988): 30.

40. See for example Johan Nordensvärd, Yuan Zhou, and Xiao Zhang, "Innovation Core, Innovation Semi-Periphery and Technology Transfer: The Case of Wind Energy Patents?" *Energy Policy* 120 (2018): 213–27; and Diego Hurtado and Pablo Souza, "Geoeconomic Uses of Global Warming: The 'Green' Technological Revolution and the Role of the Semi-Periphery," *Journal of World-Systems Research* 24, no. 1 (2018): 123–50.

41. International Federation of the Phonographic Industry (IFPI), *Global Music Report 2024: State of the Industry* (London: IFPI, 2024), https://ifpi-website-cms.s3.eu-west-2.amazonaws.com/IFPI _GMR_2024_State_of_the_Industry_666e61ca2c.pdf, 10.

42. MAHASZ, "Fizikai és digitális értékesítés 1999–2023 (millió Ft, nettó)," *Statisztikák*, accessed March 29, 2025, www.mahasz.hu/piaci_adatok.

43. Interview with a digital distributor employee, September 14, 2023.

44. Enikő Virágh and Zsolt Főző, *ProArt Zeneipari Jelentés 2018* (Budapest: ProArt, 2018), https://zeneipar.info/letoltes/proart-zeneipari-jelentes-2018.pdf.

45. Believe Amplify, November 13, 2023, Akvárium, Budapest.

46. Interview with a representative of MAHASZ, January 31, 2024. See also Tamás Tófalvy, Koltai Júlia, Knap Árpád, Mogyorósi Pálma, and Musza Stefi, *A magyar populáris zene a globális platformokon* (Budapest: NMHH, 2023), https://onlineplatformok.hu/files/b92235b0-8169-4a1c-a53e-c1bd1c6e7338. pdf. At the same time, when we view total streams, international recordings continue to dominate the market.

47. Pablo Bello and David Garcia, "Cultural Divergence in Popular Music: The Increasing Diversity of Music Consumption on Spotify across Countries," *Humanities and Social Sciences Communications* 8, no. 182 (2021), https://doi.org/10.1057/s41599-021-00855-1.

48. Interview with a digital distributor employee, September 22, 2023.

49. Interview with a musician, December 15, 2023.

50. Interview with a representative of EJI, January 2, 2024.

51. Interview with a digital distributor employee, September 14, 2023. Revenue per mille (RPM) is a metric indicating the amount of revenue earned after one thousand streams. It enables an expression of inequalities in streaming earnings.

52. Interview with a digital distributor employee, September 14, 2023.

53. Hesmondhalgh, "Is Music Streaming Bad for Musicians?"; Lee Marshall, "'Let's Keep Music Special. F—Spotify': On-Demand Streaming and the Controversy Over Artist Royalties," *Creative Industries Journal* 8, no. 2 (2015): 177–89.

54. Interview with a representative of EJI, January 2, 2024.

55. EJI, "Hungarian Performers Receive Royalties for Tens of Millions of Streams," press release, May 15, 2019, www.fim-musicians.org/wp-content/uploads/press-release-eji-deezer.pdf.

56. Qu, Hesmondhalgh, and Xiao, "Music Streaming Platforms."

57. Jason Toynbee, *Making Popular Music: Musicians, Creativity and Institutions* (London: Arnold, 2000).

58. Interview with a digital distributor employee, October 14, 2023.

59. Interview with a digital distributor employee, October 14, 2023.

60. Interview date omitted.

61. Interview with a digital distributor employee, August 28, 2023.

62. Interview with a digital distributor employee, August 28, 2023.

63. Interview with a digital distributor employee, August 28, 2023.

64. József Böröcz, "Goodness Is Elsewhere: The Rule of European Difference," *Comparative Studies in Society and History* 48, no. 1 (2006): 110–38.

65. Interview with a digital distributor employee, September 14, 2023.

66. Interview date omitted.

67. See for example Mark Banks, "Autonomy Guaranteed? Cultural Work and the 'Art–Commerce Relation,'" *Journal for Cultural Research* 14, no. 3 (2010): 251–69; David Hesmondhalgh and Sarah Baker, *Creative Labour: Media Work in Three Cultural Industries* (New York: Routledge, 2011).

68. Interview with a digital distributor employee, September 14, 2023.

69. Caplan and Gillespie, "Tiered Governance."

70. Interview with a digital distributor employee, September 14, 2023.

71. The employee here refers to Deezer's termination of its free subscription, yet, in effect, Deezer continues to be present on the Hungarian market at the time of writing.

72. Interview with a digital distributor employee, August 28, 2023.

73. Interview with a musician, December 13, 2023.

74. Interview with a musician, December 15, 2023.

75. A Hungarian popular music genre associated with weddings and similar occasions (literally: music for partying/celebrating).

76. Lofi Girl is a French record label and a hugely popular livestream YouTube channel that specializes in lo-fi hip hop music, established in 2017.

77. Interview with a musician, January 29, 2024.

78. Haynes and Marshall, "Reluctant Entrepreneurs."

79. Emília Barna, "Between Cultural Policies, Industry Structures, and the Household: A Feminist Perspective on Digitalization and Musical Careers in Hungary," *Popular Music and Society* 45, no. 1 (2022): 67–83.

4

Charting Anonymous Hits

How Short Video Platforms Have Changed the Chinese Music Industries

Shuwen Qu and D. Bondy Valdovinos Kaye

1   THE RISE OF DOUYIN AND ANONYMOUS HITS

In April 2018, a track called "Learning to Meow," accompanied by a choreographed dance, became popular on short video platforms (SVPs) and music streaming platforms (MSPs) in China. The viral explosion of "Learning to Meow," which originated on the Chinese SVP Douyin, was among the first to reveal the profound industrial and cultural impact of SVPs on the Chinese music industry. Industry reports indicate that the burgeoning short video industry is challenging the power of Chinese MSPs by cannibalizing 80 percent of the marketing and promotion budget of Chinese music labels and artists, facilitating "intermedia migration" of users from MSPs to platforms like Douyin, Kuaishou, and Bilibili, decreasing the average usage time on MSPs, and gradually slowing their revenue growth.[1] Since then, Douyin has not only evolved into one of the most powerful platforms for music promotion but has also influenced shifts in music production culture toward making what are referred to in China as "Douyin hits."

Douyin hits (*shenqu* or *baokuan*) are viral songs produced with SVP-specific logics and affordances in mind. These songs typically feature highly imagistic lyrics, memorable fifteen-second hooks, and short lifespans of three to six months.[2] Despite the massive popularity of top-charting Douyin hits, SVP users often barely recognize or remember the names of artists behind them, or sometimes even the names of the songs that go viral. This form of anonymity differs from the invisibility traditionally experienced in earlier eras by nonfeatured contributors, such as session musicians and audio engineers.

64

Rather, this anonymity is associated with a deliberate *depersonalization* of hits, where the artist's identity is intentionally de-emphasized as part of the production logic.

Against this backdrop, in this chapter we examine the problem of anonymous Douyin hits in the context of the "volatile dynamics" of platformization of music in China.[3] Previous scholars have studied the platformization of music on more music-centric platforms, like Spotify and SoundCloud, highlighting the "curatorial power" of large platforms to create playlists and the relatively limited interventions of "alternative" music platforms.[4] But little is known about the role of SVPs in the streaming music ecosystem or their interactions with other platform "complementors."[5] Scholars have also examined the various "platform effects" that encourage artists and labels to adapt certain aspects of their new releases in hopes of achieving success on platforms.[6] Although some studies suggest independent creators resist such "optimization,"[7] including in the context of playlists, there is limited research on how these dynamics play out. Moreover, no studies have yet examined these processes on SVPs, which operate outside music-centric platform systems. This chapter addresses these gaps, exploring how SVPs like Douyin interact with MSPs in shaping music production and circulation. Specifically, we ask: To what extent do promotional factors drive the production of anonymous short video hits? And how do these factors influence optimization in the music production process to increase the chances of success on both SVPs and MSPs?

In this chapter, we investigate the studios and production companies in China that focus on creating anonymous hits on the Douyin SVP. We consider in-house studios owned by or affiliated with Douyin and MSPs such as NetEase Cloud Music, QQ Music, and Kugou Music, as well as independent hit production companies, to understand Douyin-dependent optimization logics and their cultural consequences. The chapter proceeds as follows. First, we historicize the industrial and platform dimensions of intersectoral corporate control in hit-making.[8] We propose the concept "curatorial optimization" to understand the politics of hit-making in the MSP and SVP era. Second, we present a brief overview of Douyin to explain how it has disrupted the MSP-centric platform music ecosystem in China. Third, we analyze the production of "anonymous" hits by unpacking the threefold curatorial optimization process of hit-making, which downplays artist identity and de-emphasizes human involvement. This is accomplished via a close reading of media coverage, trade publications, and in-depth qualitative interviews with fifteen industry actors from Douyin, NetEase Cloud Music, hit-making studios, and licensing companies involved in Douyin hits in China. In doing so, we provide empirical evidence for understanding the implications of the audiovisual optimization of music in the streaming era, including its continuities and discontinuities with earlier industry practices.

2   CHARTS AND HITS: RETHINKING "CURATORIAL POWER" AND "OPTIMIZATION"

To understand short video hit production, we first situate cultural production for SVPs in the wider context of the platformization of music. As the term "platform" has become more ubiquitous in studies of cultural industries, there has been growing academic interest in theorizing the platformization of cultural production.[9] Some recent research has examined the platform-dependent, modular, and malleable contingency of cultural commodities that are increasingly produced for and consumed via digital platforms.[10] Douyin hits exemplify such platform-dependent, contingent cultural production. However, they are not entirely new as musical commodities, exhibiting clear continuities with earlier periods of hit-making. Examining the history of hit production in earlier recording eras reveals persistent cultural forces that shape the streaming era.

The concept of hit music in the United States emerged in the mid-twentieth century when the key trade magazine *Billboard* began tracking sales and plays on "hits charts."[11] Popularity charts are statistical evaluations of the cultural industries extending well beyond music. In his research on UK singles charts, Richard Osborne draws attention to key aspects of what we call "chart politics," including "who is responsible for compiling the singles chart, the breadth of the survey, the representation of the results, the frequency of the chart, and how a 'single' is defined," showing that "charts have never been independent of music industry stakeholders."[12] Record companies, retailers, chart companies, and music press have constantly manipulated charts to serve their interests, and there have been many debates about the degree to which charts accurately reflect public musical taste. Osborne notes that some of these concerns have diminished in the streaming era because "streaming companies such as Spotify and YouTube are the main mediators of their own hits."[13]

Such chart politics still exist in the streaming era, with concerns emerging about "curatorial power" and "cultural optimization" on platforms.[14] On MSPs, both curation and optimization involve playlists, which serve as tools for music curation and music discovery. According to Robert Prey, curatorial power refers to "the capacity to advance one's interests, and affect the interests of others, through the organizing and programming of content."[15] For Prey, this power is subject to broader structural dynamics and relations of three types of markets—music, advertising, and finance.[16] Curatorial power can be understood as a reconfiguration of promotional and distribution power. Media theorist Nicholas Garnham[17] posits that cultural distribution, rather than cultural production, is the key locus of power and profit in cultural industries. Building on his seminal work, Leslie

Meier further explored the rise of promotional industries—advertising, branding, lobbying, marketing, and public relations—as dominant forces shaping the music industry.[18] Since the 1990s, these industries have driven the "artist-brand" paradigm, wherein consumer brands and media companies "joined entrenched major music corporations as new music industry gatekeepers."[19] They compete and collaborate to forge a "cross-sector promotional apparatus," which underpins the production and marketing of popular music as brands.[20] In the streaming era, IT companies joined record companies and promotional brands as further mediators shaping hits.[21]

This brings us to "cultural optimization," a second key concept in understanding chart politics in the streaming era. Media researchers have examined how computational processes seek to make content "algorithmically recognizable,"[22] such as by refining metadata for better search-engine visibility.[23] While previous studies suggest that some music producers are inclined to resist such adjustments, pushing back against the idea of changing their final creative product to suit streaming platforms, others admit they would "devote efforts to optimization tactics such as strategizing playlist inclusion or planning a pitch for a song."[24] According to Jeremy Wade Morris, optimization logics intervene earlier and more deeply in the creative process than during the prestreaming era. Yet the legitimacy of such strategies is often contested, as it depends on negotiated interests among platform stakeholders, resulting in "definitional gaps" between acceptable and unacceptable optimization practices.[25]

In the Douyin-centric Chinese music ecosystem, however, these definitional gaps appear less evident. Douyin's immense curatorial power compels other MSPs and complementors to optimize music according to its "chart politics" (more on this below). To capture this dynamic, we propose the concept of "curatorial optimization"—a process where cultural optimization is not merely about enhancing content visibility but is instead aligned with and driven by curators' interests. Here, optimization serves the priorities of curatorial stakeholders, fundamentally reshaping the logics of hit-making.

A prominent feature of curatorial optimization is the increasing "anonymization" of artists within the Chinese pan-entertainment music industries, whereby songs have become more important than artists or albums.[26] Such anonymization is not without precedent. Dynamics of stardom and authorship have long coexisted with more anonymous production of music for commercial purposes, such as composing music for advertising or functional and ritual use.[27] Writing in the 1980s, Will Straw found that the music videos of the time dismantled existing relations between songs, albums, and the identities of performers. The "image" of individual performers became less important than the look and feel of videos.[28] Straw's insights prefigured Leslie Meier's analysis of the "possessive promotional logics" underpinning brand-music production, where songwriters are pressured to produce "sync-friendly" materials tailored to brand needs, often censoring politically sensitive content.[29] Meier further argued that an increasing reliance on consumer advertising data makes artists replaceable as long as that artist can deliver credibility and attract sufficient audiences.[30]

Arguably, however, in the streaming (and now short video) era in China, this trend of anonymization has reached unprecedented levels. Streaming platforms further depersonalize creation and marketing, stripping away much of the individual identity historically associated with performers. Before explaining how this intensified process of artist anonymization operates in the platformization of music in China, we now examine how the Chinese streaming music ecosystem has come to be dominated by tech companies and restructured under Douyin-centric curatorial power.

3   PLATFORMIZATION OF MUSIC IN CHINA AND DOUYIN'S DISRUPTION

To understand contemporary music production for SVPs in China, we must first consider Douyin's role in the platform ecosystem. Scholars have approached the platformization of music in China from various vantage points. First is the expansion of copyright. Studies have focused on the shift from an informal, peer-to-peer, and piracy-based music industry to a more formal, licensed business.[31] While widespread piracy was seen as inhibiting industrial development of cultural industries,[32] it also allowed various stakeholders to experiment with digital media, including bulletin boards, guerrilla services, and early streaming platforms.[33] This experimentation came to an end with the tightening of copyright enforcement around 2015. In the years since, the Chinese digital music industries have come to be dominated by the tech giants Baidu, Alibaba, and Tencent.[34] A second and distinct perspective focuses on infrastructural changes in China. This approach challenges views that see Chinese digital music systems as globally exceptional, emphasizing the commonalities of infrastructural design and storage media in China with those elsewhere.[35] A third approach, emphasizing transformations in Chinese cultural industries, foregrounds the distinctive music-industry ecosystem in China, with intense competition between music and other cultural platforms like Douyin and Bilibili.[36] Our focus here on "Douyin hits" follows the third approach, illuminating interactions between SVPs and the Chinese music industries, specifically via the "curatorial optimization" of music.

There are two distinctive traits of Chinese digital music that help explain Douyin's disruption of the MSP-centric platform ecosystem. First, major Chinese MSPs are part of larger tech giants that have established their own cross-sector platform ecosystems that encompass music, social entertainment, messaging, gaming, and e-commerce.[37] The economic success of Chinese MSPs relies on a "pan-entertainment" business model that involves paid subscription, virtual gifting, and tipping. The digital music business in China is, therefore, deeply embedded in cross-sector platform activities.[38] While integration within the infrastructural platform ecosystem facilitates user integration, cross-promotion, data synergy, and multiple revenue streams, it also means that Chinese MSPs are not standalone platforms to the same extent as MSPs in other regions.

Second, unlike western platforms, Chinese MSPs like Tencent Music Entertainment (TME) and NetEase Cloud Music (NCM) have been able to establish vertical integration in that they control rights to music as well as distribution (whereas rights-holding record and publishing companies in the west have so far, via their licensing agreements, prevented MSPs from holding music rights).[39] Such integration gives Chinese MSPs considerable power by bringing music, promotion, and distribution into their pan-entertainment operations. While western MSPs are still heavily reliant on licensing content from rights-holders, especially the three major western record labels, Chinese MSPs have much more leverage. For example, by 2020, TME had gained a 72.8 percent market share of Chinese music copyrights.[40]

In the wake of TME's dominance of music copyright ownership, other Chinese platforms have sought to implement similar strategies. For instance, while NCM tried to do so by incorporating self-releasing musicians into its operations,[41] Douyin has adopted a comparable approach since its launch in 2016. To reduce its reliance on licensing music from the three western major labels and TME,[42] Douyin pursued a strategy aimed at "starving the big and feeding the middle and small."[43] Between 2018 and 2020, it steadily built its repertoire by financing recordings with medium-sized copyright owners and acquiring smaller Chinese rights-holders. Qian Zhang and Keith Negus observe that "in a short period of time, Chinese platforms shifted from scrambling to sign vast numbers of musicians and their repertoire to contracting individual songs for their potential to be broken into 15-second fragments and used as sonic context in multiple short videos."[44] This observation captures an interesting shift in the cultural production of MSPs; platforms do not aim solely to amass extensive catalogs but increasingly prioritize the functional adaptability of songs for short video formats.

We argue that the shift reflects intensified competition within the vertically integrated business models pursued by Chinese tech giants. Platforms like NCM and TME are now forced to contend with the surging popularity of "Douyin hits." A key factor in this shift is that Douyin's parent company, ByteDance, has emerged as a rising tech giant, directly competing with the parent companies of MSPs, such as Tencent, NetEase, and Alibaba.[45] Revenues generated on Douyin, which mainly come from online advertising, its library of video content, and commission fees from e-commerce, position it as a powerful force in the digital space.[46] With its attention-grabbing hits, Douyin has disrupted the previously MSP-led industry, making it increasingly dependent on Douyin's influence and pushing MSPs to optimize their production according to Douyin's curatorial agenda: "to grasp more abundant data resources and absolute power in commercial bargaining and controlling visibility of video content."[47]

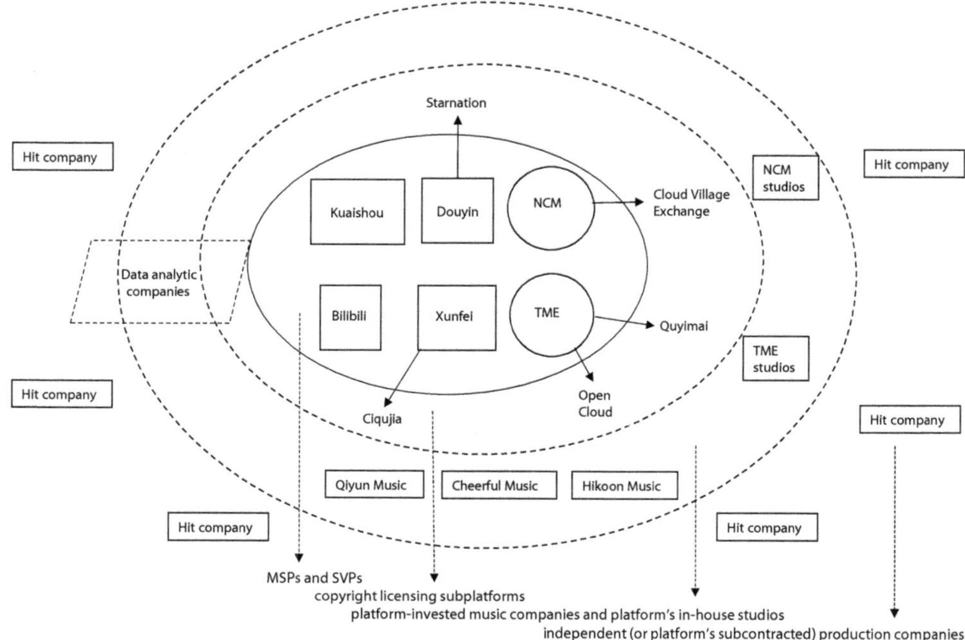

FIGURE 4.1. Major stakeholders in the Chinese hit production ecosystem. Generated by the authors.

Figure 4.1 illustrates the major stakeholders in the hit production ecosystem, including SVPs, MSPs, data analytics companies, licensing companies, and production companies. As Tatiana Cirisano notes, publishing rights are more important and can generate more revenue than recording rights for rights-holders licensing music to social entertainment services like SVPs. Publishing rights—pertaining to the ownership of musical compositions rather than sound recordings—enable the licensing of music for cover versions, remixes, and lip-synch performances. In contrast, recording rights only permit the use of tracks in their original form, prohibiting remixes or other adaptations.[48] This explains why Chinese MSPs and SVPs have heavily invested in consolidating with publishing rights-holding companies within the platform ecosystem. These platforms have launched subplatforms for content licensing, such as Quyimai and Starnation, and invested in rights-holding companies, such as Hikoon Music and Qiyun Music. These companies collaborate with subcontracted hit-making firms to acquire or buy out publishing rights for new releases. MSPs have also established in-house production studios dedicated to producing hits or scouting emerging viral tracks, securing both publishing and recording rights to further promote music. For example, NCM's studios, including Jufeng, Qingyun, and Yunshang, collaborate with approximately seven hundred subcontracted companies to optimize Douyin hits and produce chart-topping songs.

4   THE CURATORIAL OPTIMIZATION OF "ANONYMOUS" DOUYIN HITS

The term "hit" refers to a cultural product that achieves commercial success—and appears in charts listing the most popular tracks in order. Though some observers predicted that digitalization would mean a move away from the traditional reliance on hits toward a market in which profitability relies on successes further down "the long tail," hits remain central to the contemporary music industries.[49] Popularity charts, akin to traditional "hit parades," maintain high visibility on digital platforms in both the west and China. Douyin, in particular, has developed its own charting system, which uses different calculation methods from traditional charts, such as those of the US music business magazine *Billboard* or western MSPs such as Spotify.[50]

In this section, we examine the cross-platform chart politics in the production of "anonymous" hits, focusing on three key mechanisms tied to the concept of "curatorial optimization": monitoring charts, stocking index, and depersonalizing audiovisuals. We argue that these charts, whether on MSPs or SVPs, are not neutral records of success but interact with one another to serve as crucial mechanisms for producing hits. Among these, Douyin's charts hold greater curatorial power, driving trends and setting agendas that influence the practices of other stakeholders. By prioritizing the interests of the advertainment industry, Douyin's curated charts amplify its dominance while reinforcing the increasing anonymity of artists.

Monitoring Charts

Charts on MSPs are special playlists with different promotional functions. Unlike other MSP playlists that are curated on an "algo-torial" basis,[51] chart listings on platforms are algorithmically curated with little editorial intervention.[52] The dynamics of what we term "chart politics" are especially evident in the interplay between MSPs' Viral Charts and Top Charts on one side, and Douyin's Top Chart on the other. While MSPs' Original Charts and New Song Charts serve primarily as promotional tools for self-releasing musicians—highlighting the popularity of their work—they exclude songs specifically optimized for Douyin hits. These three chart types play a crucial role as tools of curatorial optimization, enabling hit studios to track trending musical content and identify potential "seed" hits— songs that exhibit early viral potential before achieving mainstream success on MSPs or Douyin.

In this process, cross-platform chart monitoring between MSPs and SVPs is fundamental for selecting "seeds." For hit-making studios, the chart monitoring process starts on the Douyin-curated Top Chart and ends at the Viral and Top

Charts on MSPs. According to one interviewee, "The Douyin Top Chart is the wind stock[53] of popularity. Only this chart can tell whether a song is popular or not."[54] This idea challenges the traditional "curatorial power" based on banner position and access to MSP playlists.[55] Due to Douyin's massive user base and traffic, its Top Chart has recentered online attention and reshaped the dynamics of music discovery and circulation. The process of identifying a promising "seed" often begins with songs appearing in the middle or lower positions of the Douyin Top Chart. Once the seed is selected, hits studios register new accounts on Bilibili[56] and Douyin to receive default recommendations. This approach avoids algorithmic filtering influenced by prior user activity, ensuring an unaltered view of the most viral content at that moment. A former employee at one studio remarked how they spent many hours monitoring data and analyzing comments of songs in the Top Chart on SVPs, despite working for NCM (which is an MSP). This NCM employee would select songs based on performance data: "If [a track] shows great momentum, like a sudden increase [in the number] of plays, we will pin down that selected song for further production and promotion."[57] This meticulous, data-driven approach underscores the strategic importance of Douyin's curatorial power in shaping the contemporary hit-making process.

Importantly, the seed being monitored is the "song" rather than the "artist." Once a seed is identified, a licensing manager approaches the holder of the publishing or song rights to either buy the rights outright or negotiate a licensing agreement. In either case, the musical artists are not considered important. What matters most to hit studios and platforms are the publishing rights of the song, not the performer's identity or the prospect of nurturing stardom. Once licensed, the seed is typically registered under the studio's name or, in some cases, a fictitious name created by the hit studio. Informants from these studios highlighted the precariousness and fragility of human artists, emphasizing that owning copyrights is far more reliable and sustainable than relying on individual performers. The seeding stage strips away performers' identity and authorship, reducing their role to an ancillary one in the broader process of hit-making.

Echoing traditional chart manipulation strategies in the preplatform era,[58] Douyin-dependent studios seek ways of calculating the popularity momentum of "seeds" by closely monitoring their shifting positions on the Douyin Top Chart and MSP Viral Charts. The MSP Viral Charts, which list the "top 100 fastest rising singles on a daily basis," provide crucial data for assessing the daily performance of these seeds. This monitoring process helps studios determine whether the seeds are steadily climbing the Viral Charts on MSPs, signaling their potential for broader popularity.[59]

Stocking Index

After a "seed" song is selected for hit-making, it enters the next stage of promotion through "stocking" the song's indices. Index-stocking is the most important stage and also the costliest stage in enabling the selected seeds to become chart hits through several rounds of index-based promotion. In the seeding stage, promising tracks often appear in unimportant charts or in the lower positions of Douyin's Top Chart, rather than in the top rankings of MSPs' Top Charts and Viral Charts, which are most important for the eventual success of hits. Osborne notes that Top 40 positions in the United Kingdom often acted "as a trigger for the resources they will commit to an act," explaining that "the chart is as an echo chamber that resounds loudest at the top."[60]

The CEO of a leading hit studio revealed that, while the costs of producing and promoting a hit were once split roughly evenly, the distribution has now shifted dramatically, with approximately 10 percent allocated to production and 90 percent to promotion. In 2023, the average promotion cost for a hit ranged from US$2,000 to US$14,000.[61] This significant increase in promotional expenses, particularly on Douyin, sharply contrasts with the shrinking cost of production. Studios allocate substantial budgets to pay Douyin influencers—including those in dance, beauty, comedy, and drama—along with promotional accounts such as news, science, information, and government channels, to feature tracks as background music (BGM) in their short videos. These efforts are supplemented by multiple iterations of digital marketing campaigns. As one informant from an NCM studio explained:

> The first batch might cost US$150 to test results from the scenery account and see how it performs. In the second batch, we'd have a meeting to decide which Douyin accounts to target next, spending a bit more to pick different ones. By the third batch, we're checking the song's ranking on the NCM Viral Chart. Our studio's key performance indicators depend on getting a top spot, as being number one will ultimately bring in new listeners.[62]

In this process, promotional activity relies on backend data analytics systems surveilling indices of songs' and musicians' popularity. "The Musician Index" is a tool to quantify the popularity of a musician based on metrics that display musicians' and listeners' behavioral, social, and relational performances.[63] Similarly, a song index is also calculated based on accumulated listeners or number of comments. Major platforms like NCM and TME have backend systems that enable employees to conduct real-time analysis. That system displays song performance data by date and time, highlighting key inflection points related to trending Douyin topics, viral videos, public events or holidays, and promotional campaigns. Informants likened that system to a "stock market," where songs function as securities. Investment in hit promotion is akin to buying shares in a rising stock: employees monitor the fluctuating performance of a song and decide the optimal moment to increase marketing expenses. Smaller hit studios also leverage data analytics by working with specialized data analytics companies or developing their own systems. Cloud Cat, a Beijing-based hit production company, attributes its success to a data analytic system "that monitors each song, intelligently analyzes the market returns of each track, and extrapolates the input-output ratio."[64]

The anonymity of a hit is accentuated at this stage. An emerging hit supported by a promotional campaign becomes a financial asset, a transactional entity quantifiable through sets of data. Index-stocking involves no engagement with or investment in the artists or performers who created the hit. Instead, promotional strategies are determined using metrics and indexing systems. This numbers-driven approach holds the potential for substantial revenue generation for hit production companies. If the track achieves one hundred million plays across several months, it can bring in millions of yuan for the hit company. As one informant said, "If you invest in one hundred songs at a time, two or three may chart. These final hits can cover all the marketing costs of the others."[65]

Depersonalizing Audiovisuals

The monitoring and stocking stages are not only vital for finding and promoting a "seed," which is already written and recorded. They are also embedded in the songwriting and recording of an original song. Emotion and affect play a central role in audiovisual optimization for short video content. Mark Andrejevic notes that in an era of information glut, affect is becoming a critical factor in cutting through the clutter and eliciting "gut instinct" responses.[66] A producer from a hit studio remarked, "To rely on a hit formula is not enough; we have to find and instill mass sentiment into that song."[67] To configure and optimize "sentimental" music, emotion needs to be translated into audiovisual languages that serve the "possessive promotional logic" of Douyin's advertising and e-commerce services.[68] In the case of Douyin hits, this process prioritizes vivid imagery and concrete depiction while simultaneously stripping away the artistic persona and traces of human involvement in production and promotion.

Studios and hit companies use various techniques to analyze sentiments expressed in user comments on MSPs and SVPs to identify themes and phrases for writing lyrics. Kugou Music, for example, displays word frequencies in comments. A producer from a smaller hit studio described how he would extract and select words from the comments of targeted songs as the thematic basis for songwriting. As one informant put it, "We write music and lyrics based on those selected keywords, like regret, loneliness, and sadness. Sometimes we even directly copy the sentences from comments to better capture and convey those emotions."[69] Additionally, they draw on news topics trending on Douyin charts, spanning areas such as social news, local news, and national news, to find inspiration for song titles and lyrics. News-based keywords are later incorporated into metadata tags, enhancing the song's discoverability and facilitating its promotion.

Besides lyrics, vocals are also extremely important for instilling affect and sentiment in hit production.[70] To translate the "aggregate sentiment" of individuals,[71] studios prioritize voices that embody what is referred to in China as a "sense of internet,"[72] rather than selecting unique voices with distinctive personalities or subjectivities. One informant remarked, "This means the voice should not remind listeners of one particular artist's face and performance, but rather ubiquitous users' faces and practices, accompanied by an anonymous voice on Douyin."[73] Vocal preference is thus not about representational accuracy, but correlational confirmation on "covert and pre-emptive opinion."[74] For instance, an ethereal female voice might be chosen to evoke the imagination of young girls wearing traditional costumes. Such decisions underscore why hits are designed not to nurture singers, songwriters, stars, or idols, but to serve as a voice-led chamber of anonymous faces in everyday life.

In terms of visuals, concrete scene-making in songwriting also plays a pivotal role in the curatorial optimization of hits, as it creates a narrative-based backdrop that resonates with users. According to an insider from Hikoon Music,[75] effective scene-making can greatly enhance a song's viral potential by fostering emotional connections, visual appeal, trend creation, and compelling storytelling.[76] The 2022 *Douyin Music Ecosystem Report* highlights the top five scenic usages of music: casual snapshots, parenting, fashion, food, and home.[77] Hit studios carefully optimize their lyrics to align with these curated scenes, leveraging trends and hashtag challenges to amplify visibility and usage. This strategic coupling improves the chances that the songs will be used in viral dance challenges and various visual scenes by Douyin influencers, Douyin promotional accounts, and ordinary users. Informants further highlighted the "fission effect" of scenic proliferation, where emerging hits evolve into unexpected, iterative versions on Douyin. For instance, the hit "Riding on a White Horse," originally performed by Xu Jiaying, has spawned multiple versions on Douyin, including DJ, campus-themed, and other remix renditions. Such scenic proliferation during hit promotional campaigns not only sustains user engagement but also drives substantial aggregate streaming and licensing revenues for hit studios.[78]

With these audiovisually optimized hits available for short video music accompaniment, brands and e-commerce services can easily incorporate music into their promotional content. Such optimized audiovisual content can boost attention and drive purchases by encouraging viewers to click on the e-commerce or advertising links attached to song information. As a video editor from a Douyin promotional account said, "We usually go for fast-paced dance tunes, and after testing, we found that the best time to drop the drums for maximum effect in product purchases is at the third or fifth second. Most of these tunes have a strong bass, usually offbeat bass. This groovy music can be nicely synced with the video edits, to stimulate purchases."[79]

Douyin's audiovisual optimization intensifies the anonymization of artists, depersonalizing music creation by utilizing data from user comments and trending topics to craft lyrics and compositions that appeal to a broad audience. Emotions and affect are meticulously analyzed and transformed into audiovisual elements that erase the artist's persona, prioritizing user engagement and maximizing viral potential. These depersonalized audiovisuals are engineered to seamlessly integrate into advertising and e-commerce, shaping a music experience tailored for commercial ecosystems, rather than artistic individuality.

5   THE PLATFORMIZATION OF POP

In this chapter, we have explored the music production culture surrounding the fascinating phenomenon of "Douyin hits" in China, as a way of investigating how SVPs are transforming the Chinese music industries, highlighting Chinese exceptionalism as part of an effort to push back against assumptions that western music business practices are global or universal. Exceptional though their industrial structure may be, Douyin hits, we argue, share continuity with elements of hit-making from the prestreaming popular music era outside of China. The making of pop hits has long been a feature of industrialized music, aimed at achieving sales, radio plays, and advertising revenue.[80] However, Douyin hits mark a departure from traditional Chinese music industry strategies, shifting toward the curatorial optimization logics of SVPs and the depersonalization of performers. By embracing Douyin-dependent hit-making since 2018, the Chinese music industries have introduced an alternative production model that challenges artist-oriented self-releasing schemes on MSPs and long-established stardom-driven practices in China and internationally.

We have focused here on the industrial production logics that undergird hit promotion, as revealed by interviews with industry actors working at hit production companies in China. Future studies should consider the ways in which the platform interfaces and affordances of SVPs shape hit production and consumption, and how these dynamics differ in international contexts.[81] Additionally, we have highlighted two key concepts central to the platformization of music production by identifying the curatorial optimization processes used to promote Douyin hits. In doing so, we have expanded beyond a close reading of MSPs and playlists to understand the ways in which the platformization of popular music operates on other kinds of cultural platforms, such as SVPs. Future research could further examine the musical consequences arising from the risk of the narrowing or disappearance of 'definitional gaps' among stakeholders in these curatorial optimization practices, as well as their broader cultural impacts on music ecosystems in the streaming era.

NOTES

1. Benjamin Walewski, "5 Things to Know about Greater China's Music Market," Believe, June 29, 2023, www.believe.com/newsroom/5-things-about-greater-china-music-market. This article describes Douyin's impressive encroachment of user base and usage time and its effect on MSPs in China. Fast-data statistics show the monthly active user base (MAU) on MSPs started to drop from 627 million users to 602 million users since 2020, and there is a consecutive year-on-year decrease rate of MAU

from 2018 to 2020. See "Fastdata 极数: 2022年中国数字音乐行业洞察报. 从音乐"卷"到直播, 涨价也救不了腾讯音乐" (Fastdata Jishu: 2022 China digital music industry insights report—From music "involution" to live streaming, price increases can't save Tencent Music), *Rongzhong Finance*, January 5, 2024, www.thecapital.com.cn/newsDetail/104653.

2. Shuaishuai Wang, "How Douyin Is Reshuffling the Chinese Music Industry," *Sixth Tone*, March 9, 2021, www.sixthtone.com/news/1006923.

3. Jose van Dijck, Thomas Poell, and Martijn de Waal, *The Platform Society: Public Values in a Connective World* (Oxford: Oxford University Press, 2018), 19.

4. Robert Prey, "Locating Power in Platformization: Music Streaming Playlists and Curatorial Power," *Social Media and Society* 6, no. 2 (2020): 1–11; Jeremy Wade Morris and Devon Powers, "Control, Curation and Musical Experience in Streaming Music Services," *Creative Industries Journal* 8, no. 2 (2015): 106–22; David Hesmondhalgh, Ellis Jones, and Andreas Rauh, "SoundCloud and Bandcamp as Alternative Music Platforms," *Social Media and Society* 5, no. 4 (2019): 1–13.

5. Van Dijck, Poell, and de Waal, "Platform Society"; Nick Srnicek, *Platform Capitalism* (Cambridge: Polity Press, 2019).

6. Jeremy Wade Morris, "Music Platforms and the Optimization of Culture," *Social Media and Society* 6, no. 3 (2020): 4–10.

7. See for example Benjamin A. Morgan, "Optimisation of Musical Distribution in Streaming Services: Third-Party Playlist Promotion and Algorithmic Logics of Distribution," in *Rethinking the Music Business: Music Contexts, Rights, Data, and COVID-19*, ed. Guy Morrow, Daniel Nordgård, and Peter Tschmuck (Cham, Switzerland: Springer International Publishing, 2022), 151–69.

8. Robyn Caplan and danah boyd, "Isomorphism through Algorithms: Institutional Dependencies in the Case of Facebook," *Big Data and Society* 5, no. 1 (2018): 1–12; David Nieborg and Anne Helmond, "The Political Economy of Facebook's Platformization in the Mobile Ecosystem: Facebook Messenger as a Platform Instance," *Media, Culture and Society* 41, no. 2 (2019): 196–218.

9. Tarleton Gillespie, "The Politics of 'Platforms,'" *New Media and Society* 12, no. 3 (2010): 347–64; David B. Nieborg and Thomas Poell, "The Platformization of Cultural Production: Theorizing the Contingent Cultural Commodity," *New Media and Society* 20, no. 11 (2018): 4275–92.

10. Chunmeizi Su, "Contingency, Precarity and Short-Video Creativity: Platformization Based Analysis of Chinese Online Screen Industry," *Television and New Media* 24, no. 2 (2023): 173–89.

11. Richard Osborne, "At the Sign of the Swingin' Symbol," in *Music by Numbers: The Use and Abuse of Statistics in the Music Industry*, ed. Richard Osborne and Dave Laing (Chicago: Intellect Books, 2021), 20–38.

12. Osborne, "At the Sign," 20.

13. Osborne, "At the Sign," 36.

14. Prey, "Locating Power in Platformization"; Morris, "Music Platforms."

15. Morris, "Music Platforms," 3.

16. Morris, "Music Platforms," 8.

17. Nicholas Garnham, *Capitalism and Communication: Global Culture and the Economics of Information* (London: Sage, 1990), 161.

18. Leslie M. Meier, *Popular Music as Promotion: Music and Branding in the Digital Age* (Cambridge: Polity Press, 2016).

19. Meier, *Popular Music as Promotion*, 84.

20. Meier, *Popular Music as Promotion*, 155–58.

21. David Hesmondhalgh and Leslie M. Meier, "What the Digitalisation of Music Tells Us about Capitalism, Culture and the Power of the Information Technology Sector," *Information, Communication and Society* 21, no. 11 (2018): 1555–70.

22. Tarleton Gillespie. *Custodians of the Internet: Platforms, Content Moderation, and the Hidden Decisions That Shape Social Media* (New Haven, CT: Yale University Press, 2018), 64.

23. Morris, "Music Platforms."

24. For examples of music creators who resisted these practices, see Morgan, "Optimisation of Musical Distribution"; Rick Everts, Erik Hitters, and Pauwke Berkers, "The Working Life of Musicians: Mapping the Work Activities and Values of Early-Career Pop Musicians in the Dutch Music Industry," *Creative Industries Journal* 15, no. 1 (2022): 97–117. For an example of those who embraced these practices, see Ignacio Siles, Amy A. Ross Arguedas, Mónica Sancho, and Ricardo Solís-Quesada, "Playing Spotify's Game: Artists' Approaches to Playlisting in Latin America," *Journal of Cultural Economy* 15 (2022): 15.

25. Morris, "Music Platforms," 4.

26. Charlie Harding, "Why Hit Songs Suddenly Matter More Than the Stars That Sing Them," *Vox*, June 21, 2021, www.vox.com/the-highlight/22538526/playlists-spotify-music-pop-stars-charts.

27. See Sesigür's chapter in this volume (chapter 11) for a study of contemporary advertising musicians in Turkey.

28. Will Straw, "Music Video in Its Contexts: Popular Music and Post-Modernism in the 1980s," *Popular Music* 7, no. 3 (1988): 247–66.

29. Meier, *Popular Music as Promotion*, 121–23.

30. Meier, *Popular Music as Promotion*, 104.

31. Jiarui Liu, "The Tough Reality of Copyright Piracy: A Case Study of the Music Industry in China," *Cardozo Arts Entertainment Law Journal* 27 (2010): 621–61; Zhen Troy Chen, *China's Music Industry Unplugged: Business Models, Copyright and Social Entrepreneurship in the Online Platform Economy* (Singapore: Palgrave Macmillan, 2021).

32. Liu, "Tough Reality of Copyright Piracy."

33. Xiaobai Shen et al., "Digital Online Music in China: A 'Laboratory' for Business Experiment," *Technological Forecasting and Social Change* 139 (2019): 235–49.

34. Shen et al., "Digital Online Music."

35. Zhongwei (Mabu) Li and David Hesmondhalgh, "From P2P to the Cloud: Music, Platformization, and Infrastructural Change in China," *Chinese Journal of Communication* (2024): 1–17; Renyi He and Jiang Chang, "More Than Just a Box: Music Storage Medium and the Mediatization of Chinese Rock Culture," *Critical Arts* 37, no. 5 (2023): 93–107.

36. Shen et al., "Digital Online Music in China"; Diming Tang and Robert Lyons, "An Ecosystem Lens: Putting China's Digital Music Industry into Focus," *Global Media and China* 1, no. 4 (2016): 350–71.

37. Tang and Lyons, "Ecosystem Lens." Notably, Apple Music and Amazon Music are exceptions; the remaining MSPs, such as Spotify, SoundCloud, Deezer, and Tidal, are not sectoral platforms.

38. Shujen Wang, "Platformization, Pan-Entertainment and Piracy: What the Fast-Changing Chinese Mediasphere Tells Us about Technology, Policy and the State," *Journal of Digital Media and Policy* 12, no. 2 (2021): 293–309.

39. Shuwen Qu, David Hesmondhalgh, and Jian Xiao, "Music Streaming Platforms and Self-Releasing Musicians: The Case of China," *Information, Communication, and Society* 26, no. 4 (2023): 699–715.

40. Xiaoling Zhou, "版权大战十年: 价格涨了近千倍, 音乐变得更好了吗?" (A decade of copyright wars: Prices have risen nearly a thousandfold, but has music improved?), *South Weekly*, August 24, 2021, https://new.qq.com/rain/a/20210824AoDKLNoo.

41. Qu, Hesmondhalgh, and Xiao, "Music Streaming Platforms."

42. Wang, "How Douyin Is Reshuffling."

43. Weida Wang, "Standing In/Out: The Platformization of Tencent's TME Live in Post-Pandemic China," *Perfect Beat* 21, no. 1 (2021): 56–62.

44. Qian Zhang and Keith Negus, "From Cultural Intermediaries to Platform Adaptors: The Transformation of Music Planning and Artist Acquisition in the Chinese Music Industry," *New Media and Society* (2024), https://doi.org/10.1177/14614448241232086.

45. D. Bondy Valdovinos Kaye, Xu Chen, and Jing Zeng, "The Co-Evolution of Two Chinese Mobile Short Video Apps: Parallel Platformization of Douyin and TikTok," *Mobile Media and Communication* 9, no. 2 (2021): 229–53.

46. Claire Fu and Daisuke Wakabayashi, "There Is No TikTok in China, but There Is Douyin. Here's What It Is," *New York Times*, April 25, 2024, www.nytimes.com/2024/04/25/business/china-tiktok-douyin.html.

47. Zongyi Zhang, "The Infrastructuralization of TikTok: Transformation, Power Relationships, and Platformization of Video Entertainment in China," *Media, Culture and Society* 43, no. 2 (2021): 233.

48. Tatiana Cirisano, "TikTok Highlights the Importance of Publishing—and Lack of Songwriter Services," MIDiA Research, February 29, 2024, https://midiaresearch.com/blog/tiktok-highlights-the-importance-of-publishing-and-lack-of-songwriter-services.

49. Anita Elberse, *Blockbusters: Hit-Making, Risk-Taking, and the Big Business of Entertainment* (New York: Henry Holt, 2013).

50. For instance, on Spotify, scrolling down the page under the "Search" column reveals the "Scan All" section, which includes "Charts" alongside other playlists like pop, country, podcasts, and audiobooks. Within "Charts," one will find Spotify Top Charts and Spotify Viral Charts.

51. Tiziano Bonini and Alessandro Gandini, "'First Week Is Editorial, Second Week Is Algorithmic': Platform Gatekeepers and the Platformization of Music Curation," *Social Media and Society* 5, no. 4 (2019): 8–11.

52. "Charts," Spotify for Artists, n.d., https://support.spotify.com/us/artists/article/charts/.

53. A colloquial term to measure performance, such as a benchmark.

54. Personal communication, 2023.

55. Prey, "Locating Power in Platformization"; Nancy Baym et al., "Making Sense of Metrics in the Music Industries," *International Journal of Communication* 15 (2021): 3418–41.

56. A longform Chinese audiovisual platform.

57. Personal communication, 2023.

58. Osborne, "At the Sign," 21.

59. From the NCM viral chart description.

60. Osborne, "At the Sign," 20–25.

61. Personal communication, 2023.

62. Personal communication, 2023.

63. Qu, Hesmondhalgh, and Xiao, "Music Streaming Platforms."

64. Xiaobai Li, "The Factory-Made Hit Song on Douyin," *Jiemian*, August 10, 2021, https://m.jiemian.com/article/6463988.html.

65. Li, "Factory-Made Hit Song."

66. Mark Andrejevic, *Infoglut: How Too Much Information Is Changing the Way We Think and Know* (New York: Routledge, 2013).

67. Personal communication, 2023.

68. Meier, *Popular Music as Promotion*, 88.

69. Personal communication, 2023.

70. Wang Yanxin, "'神曲'不是偶然，这些公司正在批量打造抖音流行" (The "hot song" is no accident, and these companies are making Douyin massively popular), *Huxiu*, September 28, 2019, www.huxiu.com/article/319929.html.

71. Andrejevic, *Infoglut*, 115.

72. Literally 网感.

73. Personal communication, 2023.

74. Andrejevic, *Infoglut*, 115.

75. Hikoon Music is a TME-invested music production company specializing in the production of popular music, especially viral hits. Hikoon works closely with TME in content creation, copyright licensing, and content distribution to strengthen Tencent's music ecosystem.

76. Personal communication, 2023.

77. *2022 Douyin Music Ecosystem Report* (Beijing: ByteDance, 2022), https://gongyi.sohu.com/a/581618980_232663.

78. Personal communication, 2023.

79. Personal communication, 2023.

80. Simon Frith, "The Industrialization of Popular Music," in *Music Sociology: Examining the Role of Music in Social Life*, ed. Sara Towe Horsfall, Jan-Martijn Meij, and Meghan D. Probstfield (Boulder, CO: Paradigm Publishers, 2013), 223–31.

81. D. Bondy Valdovinos Kaye, Jing Zeng, and Patrik Wikström, *TikTok: Creativity and Culture in Short Video* (Cambridge: Polity Press, 2022).

5

"From the Region, For the Region"

Anghami and the Postcolonial Challenges of Localizing Music Streaming in Emerging Markets

Darci Sprengel

Anghami, meaning "my tunes" in Arabic, is a music streaming service founded in Beirut, Lebanon, in 2012. Using the slogan "From the region, for the region," it promises to better meet the needs of Arab listeners than its international rivals. Boasting a catalog of fifty-seven million songs and seventy million users, it claims to be "revolutionizing digital music consumption" by combining international and local sounds in ways it asserts listeners in the MENA/SWANA (Middle East and North Africa / Southwest Asia and North Africa) and its diaspora want to hear. According to the service's own data, it is preferred not only across the MENA/SWANA region but also within the Arab diaspora, especially among new migrants to Europe who continue to use the platform because, as one of the service's founders said, "Anghami reminds them of the 'scent' of their home, of their streets in the Middle East."[1] In short, Anghami's distinctly local knowledge is its "sonic brand."

But Anghami isn't the only music streaming player in the market claiming to go local. In September 2018, Spotify launched its Global Cultures Initiative, which it insisted was one of the "most important things," making it a "leader" in the field of audio streaming by moving the platform beyond its traditional focus on North American and European musics to "promote and advance culturally diverse music."[2] Between 2018 and 2024, Spotify expanded to over 150 new markets, including the MENA/SWANA region. Claudius Boller, former managing director of Spotify Middle East and Africa, said of this expansion, "Spotify's ambition is really to bring music to everyone, and we need to be 100% locally relevant with our consumer offering."[3]

Such claims suggest the importance of a rhetoric of global "diversity" as music streaming develops as a global technology. For both Anghami and Spotify, notions of "local knowledge" and "cultural specificity"—in general, a concept of culture and of cultural difference—are at the forefront of these considerations. This raises questions regarding how multinational streaming services employ ideas of culture and the local, and to what ends. In short, how do music streaming services define the "local"? What work is a concept of "culture" doing here?

This chapter demonstrates that Anghami navigates a difficult position in the field of music streaming. As the platform contends with resilient colonial narratives of Western superiority in the realm of technological development, it cannot define local music streaming entirely on its own terms, but must instead situate the service in relation to regional ideas about what constitutes global culture. This at times manifests as points of distinction—for instance, to differentiate Anghami from its non-Arab competition—and at other times in terms of sameness, especially to align with capitalist values and its logic of a blockbuster model in the music industry. That the platform expresses these logics of difference and sameness simultaneously suggests a particular view of the local that aligns with the logics of global capitalism. In line with these logics, Anghami foregrounds its racial/ethnic distinctiveness as a means to secure a foothold in a saturated streaming market, potentially repelling domestic users who harbor an aversion toward what is deemed too "local." At the same time, it reproduces the dominant for-profit, subscription-based model of music streaming espoused by Spotify, which often disadvantages more musically niche artists in the region. Overall, I suggest that as Anghami attempts to cater its service to a MENA/SWANA user base, it navigates a particular postcolonial condition wherein it reproduces certain capitalist logics of accumulation, dominant in the global music industry, in order to participate in it while also striving to address the needs of listeners and musicians back home. Put simply, it cannot be too local or too global.

Considering the relationship between streaming services and ideas of the local is critical, given the music industry's problematic history in perpetuating essentialized tropes of "cultural difference" to market non-Western music. As ethnomusicological analysis of the world music industry has demonstrated, these marketing techniques rely on racialized tropes that exacerbate the unequal treatment of non-Western artists. And as David Novak and Kyra Gaunt have shown, new technologies such as the internet largely continue to perpetuate inequalities along lines of race, ethnicity, and gender.[4] But Anghami's marketing techniques target listeners in the MENA/SWANA region, not the predominantly white, middle-class Western subjects of world music literature—so what work is a concept of local culture doing in this case, and does it transform these longer debates?

This chapter is based on over fifteen years of ethnographic research in Egypt, with a particular focus on one year of fieldwork in the United Arab Emirates and Egypt between 2023 and 2024. During this year, I conducted ethnographic interviews with around sixty industry professionals and musicians in the Arabic music industry, along with about twenty users of Spotify and Anghami. I also engaged in participant observation at various concerts and industry events. Interviews were held in English or Arabic, depending on the interviewee's preference.

THE SHIFTING STATUS OF THE LOCAL IN THE MUSIC INDUSTRY

Analyzing the recent global expansion of Spotify and Netflix, Evan Elkins argues that globalizing digital platforms attempt to soothe anxieties about their imperialistic economic dominance by framing their globally expanding businesses as benevolent humanistic and cosmopolitan technologies that facilitate cross-cultural connection and global community.[5] Drawing from Anna Tsing's important work, he asks, "What vision of the globe are they offering, and how do they present themselves as the ideal institutions to help encourage this vision?"[6] Part of the way these platforms present themselves as global institutions is by removing context. Spotify is not merely a Swedish technology, nor is Netflix exclusively American. Instead, they are global, universal entities, defined not only through their geographic expansion and financial dealings but also textually through their rhetoric and discourse. And yet, they still brand themselves as caring deeply about cultural specificity and encouraging cross-cultural connection.[7] Projects like Spotify's Global Cultures Initiative thus demonstrate how streaming services position themselves as both universal and global technologies, decontextualized from any community of origin, while *simultaneously* presenting themselves as deeply localized or localizable, tailored in each instance to particular listeners, cultures, and places.

Although the globalizing aspects of streaming technologies have received some scholarly attention, the role of the local and of localization remains less understood to date. Looking at music streaming, scholars have noted that its advent around the globe can result in the transformation of local music cultures. Shuwen Qu, David Hesmondhalgh, and Jian Xiao argue that, in China, music streaming has facilitated the incorporation of previously independent musical activity into the business models of the music industry, offering self-releasing musicians new avenues for making money while also constricting their autonomy in ways that reaffirm the industry's centralization.[8] Likewise, in India, platformization has challenged the long-standing domination of Indian film music and accelerated efforts toward copyright reform, benefiting some non-film musicians who previously led highly precarious careers, while leaving others behind.[9] Such studies demonstrate how streaming technologies are always connected to global digital developments while also reflecting their own specificities and logics as they interact with local music practices and histories in each context.[10] They show that streaming services, rather than simply responding to or empowering local practices neutrally, actively intervene in music scenes and cultures, transforming practices and hierarchies while reinforcing existing power structures—often at the same time.

The rhetoric of music streaming services, as at once "global" and "local", broadly signals a transformation in the status of the local within the dominant music industry. For instance, although the majors in the Western recorded music industry have long held regional offices around the globe, these offices primarily focused on marketing existing Western music catalogs to global audiences, doing little to support local music and musicians in their respective locations.[11] They treated global markets as consumers rather than producers of the dominant music industry's products. Music streaming services such as Spotify and Anghami, by contrast, claim to champion local music production and, in so doing, democratize the music industry by shifting power away from global centers. For instance, Spotify's managing director for the MENA region and South Asia, Akshat Harbola, said that the service's priority is "discovery": "We want to provide more opportunities for artists from this region to be discovered and enable the consumption of music from the region, both locally and globally."[12] Such corporate rhetoric that champions non-Western music markets as global music producers invites us to reformulate Elkins's question: What vision of the *local* are these platforms offering, how do they present themselves as ideal institutions to engage local listeners and musicians, and how might diverse listeners and musicians respond and speak back to these technologies?

Significantly, streaming's repositioning of the status of the local is not new in the music industry. Scholars in the 1990s characterized the rapid globalization that occurred over the 1970s and '80s as having transformed the status of the local. Ethnomusicologist Jocelyne Guilbault, for example, argued at the time that it was in the interest of those within the dominant market to (re)define the local through the category of "world music," due to fears that diversification and fragmentation would mean losing monopolistic control. The potential "ethnicization of the mainstream" brought by globalization meant that the privileged position of those in global centers of power could be challenged to such an extent that they no longer defined global culture. Those in "emerging" markets likewise had an interest in the world music label. For them, it was a way of (re)defining the local to protect against losing cultural identity to perceived worldwide homogenization or, alternatively, an opportunity to promote cultural identity. Both protecting and promoting local culture were subversive responses and ways of participating more actively in the international market.[13] The "world music" category was thus one avenue through which emerging markets became part of the dominant music industry, reflecting "the desire of every nation not only to be recognized but also to *participate* in the workings of global economics and power."[14] In this way, the dominant Western music industry could regard non-Western markets as music producers, but only under a homogenizing world music label that catered to the tastes and marketing needs of Western audiences, thereby diluting local meanings, understandings, and relations of power in the process. This discourse took place within the larger context of world record sales being dominated by a few Western countries, while non-Western music was being marginalized to stations outside of mainstream radio.[15] For other scholars, the category of world music proved that the music industry could no longer be conceived as a bilateral, center-periphery market. Debates around hybridity sought to go beyond this neocolonial binary while acknowledging its historical significance.[16] With the development of music streaming technologies, the meaning of the local may once again be shifting, and in ways that are not entirely novel. What are the terms through which regional entities can operate in a more global music economy, and are they the same terms as before? How well suited are notions such as appropriation, hybridity, and center-periphery for understanding streaming services developed by and for communities within the Global South?

As a homegrown music streaming service "by and for the region," Anghami is significant for challenging a history of Western dominance and control over music technologies. For instance, when record companies first appeared in Egypt in the early 1900s, they were all foreign-owned. Records were pressed in Europe and imported to Egypt, increasing the cost of these products considerably. It wasn't until fifty years later that Egypt established its first record manufacturing factories, but even then, they still relied on foreign "expertise" and equipment.[17] As ethnomusicologist Michael Frishkopf writes, "Egyptian composers, lyricists, singers, musicians, and producers all suffered from foreign domination and exploitation of local music production," even while the Egyptian film and music industries thrived in the production of local content.[18] In many ways, streaming services designed by and for non-Arabic speakers continue this dynamic. Technological development and production are made in a geographic elsewhere, designed with non-Arabic music and speakers in mind, thereby perpetuating a relationship of dependence on foreign technological creativity, innovation, and expertise.

By contrast, Anghami offers the region's listeners and musicians the chance to support a sonic technology specifically made for Arabic music and speakers—a potentially subversive act. As Sarah El Miniawy, founder of Simsara Music, a management company focusing on innovative Arabic independent music, told a panel at the Ma3azef Symposium on Arabic Music at the Sharjah Art Foundation in 2019, "I love Anghami because it is homegrown." Anghami users told me that they preferred to use the app over Spotify because Anghami had a more extensive catalog of Arabic music and did a better job curating Arabic content through its many playlists organized around location, nation, time period, mood, genre, and so on. Anghami users can also toggle their settings to display only Arabic results, only international, or both Arabic and international. For the Anghami users I spoke to, this better represented how they preferred to listen to music. It helped ensure that the Arabic content on the platform did not get drowned out by international content, unless that is what the listener wanted. In short, rather than offering a view of Arabic music catered to non-Arabic speaking listeners—an accusation charged at Spotify—Anghami was largely understood by its users as having a deeper understanding of Arabic music and presenting it more effectively to listeners who were already knowledgeable about these genres.[19] In so doing, it avoids some aspects (though not others) of dependence on foreign music industries and their technologies for local/regional music dissemination, keeping music revenue and technological development more firmly in the region.

The centrality of the local to Anghami's branding was best demonstrated when one Anghami employee sent me a report that detailed the platform's regional insights. The study was conducted in early 2020 by Ipsos, a multinational market research firm based in Paris, which examined music streaming use in Saudi Arabia, the UAE, and Egypt. The stated aim of the research was to "understand brand affinity, to identify emotional connectors, and to uncover the distinctive traits that enable a brand to win the hearts of its listeners." It claimed that its quantitative methodology enabled "in-depth analysis of consumer sentiment going beyond a simple examination of listener behaviour."[20]

The report reached two broad conclusions: first, that the MENA/SWANA region, and Saudi Arabia especially, was an area of tremendous growth potential for music streaming; and second, that the key to brand loyalty is "affinity" and the key to affinity is "local culture." Put differently, the key to growth for streaming services is utilizing consumer's affective attachments toward local culture. This simple claim was repeated ad nauseum throughout the report. For example, it concluded that

> the ease of accessibility, the availability of preferred music, and the large variety of Arabic and local music are the top three reasons why streamers use Anghami. A homegrown brand, it has remained resolutely focused on the MENA region and has developed an in-depth understanding of its cultural nuances. It is for these reasons that Anghami excels at local content and is favoured for its delivery of exclusive content.

The report further states that building emotional connections with audiences is key to effective branding. What makes a music streaming service successful is its ability to cultivate affinity for one's local culture; therefore, a company's regional origin is crucial in creating brand loyalty through trust and authenticity. Local platforms

> have humanised their product by recognising their audience's identity and culture, connecting with it, and augmenting that identity with the adoption of local talent and the provision of unique content. Not only is Anghami viewed as the most trustworthy brand amongst music streaming platforms, it is also viewed as the best when it comes to recognising Arab users' identity. An identity that is not based solely on language, but also on culture, traditions and faith.

As if this point wasn't yet driven home, the report later states, in all capital letters, "music streaming preferences are driven by how local a brand is."

This perspective positions the local as *the* prized characteristic within the music industry. Unlike the terms of participation that characterized emerging markets during the era of world music in the 1980s and '90s, this vision of the local is not a marketing category aimed at Western audiences; rather, it is specifically geared at Arabic speakers. Indeed, Anghami, as a word, is incomprehensible to non-Arabic speakers. As a translation of "my tunes," Anghami still animates a form of hybridity; however, this hybridity is only evident to those fluent in both Arabic and English. Significantly, this approach to championing the local is not exclusive to Anghami; major entities in the dominant Western music industry, such as Spotify, also employ this rhetoric. The status of the local is thus shifting in the music industry from a marketing category primarily aimed at Western audiences to one designed to attract non-Western audiences and music producers as well.

WHO REPRESENTS THE REGION?
SLIPPAGES BETWEEN "TOO" LOCAL AND "TOO" GLOBAL

But how does Anghami define the local, and how does the service represent it? Considering that the MENA/SWANA comprises twenty-two nations, each with distinct Arabic dialects, cultures, and histories, there are real challenges to including the full diversity of the region. Anghami champions ideas of the local both on and off the app. For example, it signs exclusive deals with regional artists, such as Amr Diab, an Egyptian pop star who has dominated the region's recorded music industry since the 1980s. During the month of Ramadan, Anghami offers special features that indicate prayer times and filter out all music, since some listeners in the region only want to consume religious content during the holy month. Additionally, and arguably more than any other streaming service in the region, Anghami offers on-the-ground services in the form of events, venues, and concerts. Since 2022, for example, it has run Anghami Lab, an impressive rooftop nonalcoholic bar and concert venue in Boulevard Riyadh City, as well as the "Beat the Heat" annual indoor summer music festival in Dubai, both of which feature artists from the region. Elsewhere, in interviews and media blasts, Anghami claims that its local values and knowledge are demonstrated in part by its large catalog of Arabic content, including both music and podcasts. For instance, in the case of Arabic independent music, Anghami has between twenty and thirty different platform-curated playlists, organized by MENA/SWANA country and other categories such as mood or genre. Some of these categories are highly specific to local events and ways of life in particular places. For example, one playlist is named "على كوبري ٧ أكتوبر (On 6 October Bridge)" and is dedicated to one of Cairo's major roads, which connects the eastern and western parts of the city and is famous for its traffic congestion. Until recently, Spotify, by contrast, had only one platform-curated playlist devoted to independent music from the region but hundreds—possibly thousands—of user-generated ones.

In so doing, Anghami presents itself as a pan-Arab platform, a position confirmed by its founders. Commenting on Anghami's moving its headquarters from Beirut to Abu Dhabi in 2021, cofounder and CEO Eddy Maroun told a reporter:

> We always had a vision for the whole Middle East and North Africa region—that's why we also have offices in Dubai, Cairo and Riyadh. . . . The Abu Dhabi headquarters move is in line with our vision to grow more into a pan-Arab platform.[21]

Originating in the nineteenth century and reaching its peak in the 1950s and '60s after decolonization, pan-Arabism is an ideology that advocates for the political, cultural, and socioeconomic unity of Arab peoples. But, rather than a single market, the Arabic music industry is better characterized as multiple smaller markets. Music industry professionals based in the region primarily divided it into four main music markets: the Levant (Lebanon, Palestine, Jordan, Syria), the Maghreb (Morocco, Algeria, Tunisia, Libya), Egypt, and the Gulf. But even within these smaller markets, some locations are more prominent than others. Due in part to the region's uneven history of colonialism, Beirut and Cairo developed robust recorded music industries, while other locations did not. One executive of a record label working out of Dubai told me, for instance, "We don't consider the UAE part of the Middle East because it is mostly foreigners." In the Gulf, Saudi Arabia takes a prominent position, due to its sizable population, the population's disposable income, and the state's recent shift in attitude toward music as a form of economic development. By contrast, the cultural production of Iraq, Sudan, Yemen, and Mauritania were never discussed by the industry professionals I spoke to, and they do not figure prominently on the Anghami app. Among other challenges, this fragmentation contributes to listening bubbles. Most listeners in North Africa and the Levant, for instance, do not grow up listening to music from the Gulf. As one Lebanese Anghami employee told me, music from the Gulf has distinct rhythms and "sounds strange to our ears."

Anghami thus defines the local as at once national, subregional, and regional (e.g., its "Top MENA Hits" playlist). In the music industry, this broader notion of the local must be invented. Pan-Arabism as a listening practice is not a given but rather something that must be cultivated. Part of the way this is done is through foregrounding this identity as a point of distinction from foreign brands. For example, *being* an Arab streaming service is central to Anghami's branding. Anghami treats Arab culture as unknowable and inaccessible to outsiders, positioning itself as the sole entity with privileged access to it. Notably, then, what makes Anghami unique is less its pioneering technologies or unique approach to the streaming industry but rather its racial and cultural identity. This would suggest that the terms for entering the global market are still reminiscent of the colonial era: it relies on a bifurcation of the world between us and them, East and West, wherein the East performs its identity as difference to gain access to global participation.

In addition to market fragmentation, class disparities also play an important role in how and to what extent Anghami is understood to represent the region. Despite its headquarters moving to the Gulf, the vast majority of Anghami's employees remain Lebanese graduates from top Lebanese universities. While chatting online via video call with a senior Lebanese employee at Anghami in 2020, I got a rare glimpse inside their home in the UAE. In the background, I saw the floor-to-ceiling windows of a high-rise apartment and a Peloton stationary bike that costs nearly US$2,000. As we discussed our shared love for cycling, I briefly saw a woman, likely hired help, enter the frame to clean. This employee earned their bachelor's degree and MBA from prestigious American universities and previously worked for a major social media platform in London.

They positioned themself as representing a local voice, but they are not necessarily intimately connected to listeners and musicians in Egypt, many of whom make only a fraction of a standard UAE salary. Commenting on the dominance of Lebanese working in the music industry out of the Gulf, an Egyptian music industry professional based in Cairo told me, "The Lebanese working in the Gulf may have our data but that doesn't mean they know our culture. They know nothing about the music scene here." Only those from the region with certain language and technical skills can immigrate to the Gulf. This executive's image of what constitutes the local shapes the platform and could exacerbate—or challenge—these existing class and cultural representation disparities in the region.

Seeming to support Anghami's claims that being "from the region" is essential, Spotify's Global Cultures Initiative was short-lived. According to the platform, Spotify disbanded the group only one year later after realizing that smaller, on-the-ground teams in local markets were more effective than one large operation based in New York.[22] In other words, it affirmed that *being* local matters in music streaming. According to one executive I spoke to working out of Spotify's Dubai office, the MENA/SWANA is a particularly tricky region to tackle remotely because of its stark economic disparities. The region includes both "extremely wealthy nations, especially in the Gulf, and extremely poor ones in the same network," requiring different strategies for each. Many of Spotify's integration features, such as those for cars and PlayStation, which are designed to achieve ubiquity, do not work as well among populations with limited disposable income.[23] In Egypt, for instance, many do not own their own cars, and even fewer have newer cars with this tech capacity. My primary contact at Spotify, a user researcher on the Growth Regions team, admitted that local knowledge is a problem for the platform, and she is personally advocating among her colleagues to utilize ethnographic and qualitative methods to better address these challenges.

Yet, Anghami's own data indicates that Spotify is trailing in the MENA/SWANA by only a few percentage points. This suggests that the "global" is also a powerful brand. Among my social groups in Egypt, many actually preferred Spotify because they primarily listened to international music.[24] Both they and the Anghami enthusiasts I interviewed felt that Spotify's algorithm was better than Anghami's at sorting international music.[25] Additionally, many musicians in the Egyptian music scene did not trust Anghami. They associated Anghami with unethical local business practices, where operations were perceived as neither fair nor conducted by the book. For example, many believed that Anghami did not pay its artists unless they were top performers. The general assumption was that Spotify, as a major foreign, international company, was more "fair." Spotify was likewise seen among some musicians as more prestigious. Reaching higher numbers of streams on Spotify was more meaningful for artists chasing a foreign audience, as the platform was believed to have a greater reach among international listeners. Some booking agents and recording executives in the region only looked at an artist's streaming numbers on Spotify as an indicator of success. As one industry executive working in the region told me, Spotify—not Anghami—has become akin to a CV. Its data indicated whether an artist was worth booking or managing. These practices reinforce a long tradition in the region of artists needing to first catch the attention of foreign audiences to gain legitimacy and garner attention back home.[26]

Disaffection with the global positioning of one's own culture, and distrust of one's own music industry and/or government, meant that some in the region preferred the global as a brand. It was in many ways more prestigious and trustworthy. Writing on the dominance of foreign powers in the region's technology, preeminent sociologist Nagla Rizk, a professor at the American University in Cairo and herself Egyptian, argues that what she calls "self-Orientalization"—"the process of internalizing Western perceptions of the region by its people"—is "one of the biggest challenges facing those who aim to voice narratives of innovation in MENA that constitute an alternative to the postcolonial ones."[27] In part, the resilience of colonial narratives of Western superiority work against Anghami's branding of "being" local. For some in the region, it is "too" local to be trusted.

In addition to branding itself as only a local technology, then, Anghami also positions itself as a multinational global company and tech producer that brings Arab identity and culture to the world. For example, beyond its many partnerships with international brands such as Amazon Alexa and Sony Music, Anghami boasted of becoming "the first Arab tech company to be listed on the Nasdaq."[28] For some professionals working in the region's music industry, this was a great source of pride. In a panel on "The Digital Content Age" at the 2023 edition of GITEX, a major annual technology conference held in Dubai, moderator Ayat Amr, for example, turned to panelist Elie Habib, Anghami cofounder and CTO, and said:

> I am so proud of the great work that you did at Anghami—being the first company listed on the Nasdaq. For me, as an Arab, for someone from the Middle East, this

really gives us hope that we can create things—*that we can innovate things to the world*. (Emphasis mine)

Such proud sentiments indicate that it can mean a great deal locally to be global. Yet most industry professionals I spoke to in both Egypt and the UAE told me that Anghami's Nasdaq bid was done "too early" and had been a "bad idea"—the value of an Anghami share plunged from US$9.70 on opening day to just sixty-one cents at the time of this writing in early 2025. Some told me that Anghami should have opened on the local stock market in the UAE first because Anghami is "too local" to be attractive to international investors. In 2023, for example, Anghami made an exclusive streaming deal with Amr Diab, one of the region's biggest pop stars, for a rumored US$10 million. According to one Western investor in media technologies in the region, a regional star like Amr Diab is not legible to non-Arabic speaking investors outside the MENA/SWANA. Such a deal thus does little to excite foreign investors trading on the Nasdaq, who are primarily concerned with Anghami's growth potential outside the region.

Some of the region's musicians, moreover, have been skeptical of Anghami's claim to be "from the region, for the region." For these critics, Anghami is too in line with the global logics of music streaming to truly benefit the region's artists. In April 2019, for example, the digital Arabic-language music journal *Ma3azef* put on a major symposium on Arab music in Sharjah.[29] One panel, "The Music Industry in Alternative and Independent Scenes," featured five independent musicians and music managers from across the region, along with Rami Zeidan, Anghami's product director at the time. Despite the moderator's best efforts to steer the conversation toward various other topics, the nearly two-hour discussion repeatedly devolved into pointed criticism of Anghami, which the outnumbered Zeidan struggled to defend. I paraphrase some of this discussion at length, maintaining the specific language used by the discussants without questioning its accuracy, because it demonstrates (a) how MENA/SWANA-based musicians understand Anghami and critique its claims of being local and (b) how Anghami's rhetoric surrounding this critique slips easily between ideas of being simultaneously global and local.[30]

Moderator Hala Mustafa opened the discussion with a question toward Zeidan about the role of the internet in benefiting independent musicians financially. Zeidan responded that revenues are distributed as a proportion of overall number of streams. It is split fifty-fifty between the artist and the platform. Half goes to Anghami, and then artists *share* the remaining half based on their respective share of the aggregate number of streams. Zeidan touched on several other points, including Anghami's distinct commitment to promoting regional artists.

But the musicians sharing the stage disagreed. Artists did not share fifty-fifty with Anghami; they shared a portion of 50 percent based on the artist's number of streams. Independent musician Abass El-Hage interjected that this made it impossible for regional artists to compete with the Drakes and Kanyes of the world: "We will be less than 1 percent of all their streams, and thus their share of the profit." Zeidan responded by immediately shifting from a discourse of the local to one of the global: "That is a very good point, but this isn't just Anghami; this is the worldwide formula for all streaming platforms." Seeming to struggle in Arabic, he switched to English to spout statistic after statistic:

> If we want to take this conversation even further into the economics of music streaming . . . the pay per stream is x. . . . Today, the economics of streaming is tight, worldwide, on any platform. We don't make money. But the economics of music streaming is booming today. Globally streaming grew 9.4 percent in the last year. Music streaming today contributes to 43 percent of music revenue. On a broader level, there are 1.6 billion people who stream music. 1.3 billion of them are on video platforms, and they consume 50 percent of the streams that get consumed. The remaining three hundred million on music streaming services consume the other 50 percent. Here is a bigger statistic: the three hundred million people who stream music from music streaming services contribute to 80 percent of the music revenue, while the remaining 1.3 billion who are on the likes of YouTube contribute the [rest].

Shortly after this point, Sarah El Miniawy responded in Arabic:

> I appreciate all the numbers, but this only works for an artist that is making beyond millions and millions of streams. . . . The whole idea that the platform takes 50 percent, that's also a bit too much in my view, because the same mechanics that apply for artists such as Beyoncé are the same ones that apply to underground/alternative artists. It doesn't make any sense. I love Anghami because it is homegrown, but, on the whole, streaming is bad for alternative music . . . [*everyone tries to interject*] For me, as an alternative artist, I made more money out of Bandcamp in eighteen months in comparison to what I will make in twenty years' time out of streaming. Because I am niche. Streaming does not work for niche. . . . It is a bit capitalist. It is a bit like taxing the rich the same way you tax the poor.

The musicians begin discussing how much they like Bandcamp before the moderator breaks in, trying to steer the conversation toward another topic. But musician Ahmed Zaighmouri refused (I am paraphrasing):

> No. First, I want to address this point that sounds like a dream statement "From the region, for the region." . . . Drake is from Canada. . . . He has nothing to do with us here. Why aren't we a priority in the region? Change the mechanism to work for us. If Anghami prioritized the artists from here, it would encourage us. The phrase shouldn't just be a vision statement; it should actually do something for the artists here.

Zaighmouri and the other artists on stage reveal Anghami's rhetoric of locality to be in some ways a marketing stunt. For Anghami to be truly local, the service's mechanism that determines which song will be recommended next and the

12. Neesha Salian, "Spotify's Akshat Harbola Is Keen to Support Local Talent," *Gulf Business*, February 15, 2024, https://gulfbusiness.com/interview-akshat-harbola-md-mena-sa-spotify/.

13. Jocelyne Guilbault, "On Redefining the 'Local' through World Music," *World of Music* 35, no. 2 (1993): 34–40.

14. Guilbault, "On Redefining the 'Local,'" 42–43.

15. Guilbault, "On Redefining the 'Local,'" 41–42.

16. Guilbault, "On Redefining the 'Local,'" 39.

17. A. J. Racy, "Musical Change and Commercial Recording in Egypt" (PhD diss., University of Illinois at Urbana–Champaign, 1977), 87–99; Michael Frishkopf, "Nationalism, Nationalization, and the Egyptian Music Industry: Muhammed Fawzy, Misrphon, and Sawt al-Qahira," *Asian Music* 39, no. 2 (2008): 35–42.

18. Frishkopf, "Nationalism," 30–34.

19. Darci Sprengel, "Imperial Lag: Some Spatial-Temporal Politics of Music Streaming's Global Expansion," *Communication, Culture and Critique* 16 (2023): 243–49.

20. Choueiri Group and Anghami, "Why It Matters to Be Local: Music Streaming in MENA," accessed March 6, 2025, https://dms.eu-central-1.linodeobjects.com/research/file/1606374518_f428c 4ef4fd41c0eb601568a8bd9b9ef.pdf.

21. Quoted in Kareem Cheheyab, "Music Service Anghami Moving from Crisis-Hit Lebanon to UAE," *Al Jazeera*, January 27, 2021, www.aljazeera.com/economy/2021/1/27/music-service-anghami -moving-from-crisis-hit-lebanon-to-uae.

22. Elias Leight, "Spotify Pivots on Global Cultures Initiative, Alarming Music Industry," *Rolling Stone*, October 4, 2019, www.rollingstone.com/pro/features/spotify-global-cultures-initiative-892762/.

23. See Khalil and Zayani, "Digitality," 175.

24. This group primarily consists of middle-class, college-educated individuals living in either Cairo or Alexandria and ranging in age from their mid-thirties to early forties.

25. I interviewed about twenty Anghami users in Cairo, Alexandria, Dubai, and Abu Dhabi in 2023. Interview participants were between twenty-two and fifty-four years old.

26. Katherine Zirbel, "Playing It Both Ways: Local Egyptian Performers between Regional Identity and International Markets," in *Mass Mediations: New Approaches to Popular Culture in the Middle East and Beyond*, ed. Walter Armbrust (Berkeley: University of California Press, 2000), 120–45.

27. See Kanta Dihal, Tomasz Hollanek, Nagla Rizk, Nadine Weheba, and Stephen Cave, "AI Oasis? Imagining Intelligent Machines in the Middle East and North Africa," in *Imagining AI: How the World Sees Intelligent Machines*, ed. Stephen Cave and Kanta Dihal (Oxford: Oxford University Press, 2023), 278.

28. Rizwan Choudhury, "This Indigenous Music Streaming Service from the Middle East Is Confident of Competing with Apple and Spotify: Report," *Mashable Middle East*, December 13, 2022, https://me.mashable.com/tech/22971/this-indigenous-music-streaming-service-from-the-middle -east-is-confident-of-competing-with-apple-an.

29. Details available at https://ma3azef.com/.

30. This approach foregrounds "folk theories" of music streaming, which are "intuitive, informal theories that individuals develop to explain the outcomes, effects, or consequences of technological systems, which guide reactions to and behavior towards said systems." See Michael Ann Devito, Darren Gergle, and Jeremy Birnholtz, "'Algorithms Ruin Everything': #RIPTwitter, Folk Theories, and Resistance to Algorithmic Change in Social Media," in *CHI '17: Proceedings of the 2017 CHI Conference on Human Factors in Computing Systems* (New York: Association for Computing Machinery, 2017), 3163–74, quoted in Ignacio Siles, Andrés Segura-Castillo, Ricardo Solís, and Mónica Sancho, "Folk Theories of Algorithmic Recommendations on Spotify: Enacting Data Assemblages in the Global South," *Big Data and Society* 7, no. 1, (2020), https://doi.org/10.1177/2053951720923377.

31. See also Khalil and Zayani, "Digitality," 173.

6

How Streaming Is Reshaping
Latin American Music Culture

The Case of Mexican Corridos Tumbados

Rodrigo Gómez, J. Ignacio Gallego, and Argelia Muñoz-Larroa

INTRODUCTION: THE SUCCESS OF MUSIC STREAMING IN LATIN AMERICA— AND MEXICO AS A KEY HUB

In the dynamic landscape of global music culture, Latin America has emerged as a vibrant hub, pulsating with a diversity of rich musical traditions. With the advent of music streaming platforms (MSPs), the region has witnessed a digital music revolution based on international connected consumption, catapulting its music onto the global stage as never before. Today, music streaming in Latin America represents not only a continuation of the region's rich musical heritage but also a beacon of innovation. From the infectious beats of salsa and samba, to the soul-stirring melodies of bolero and bossa nova, and the provocative mix of corridos tumbados and reggaeton, the diverse array of musical genres in the region has become increasingly attractive to international audiences.

As Hesmondhalgh remarks, "The massive new role of Music Streaming Platforms (MSPs) in musical consumption means that they increasingly operate as the core of the music industries and of the everyday experience of recorded music."[1] This is certainly true of Latin America. Streaming music consumption in that region, which has a population of roughly 660 million people, has been steadily increasing due to improved internet connectivity, smartphone penetration, and the availability of affordable, primarily free streaming services. Tech companies monetize these cultural consumption offerings through various means. In the Latin American context, streaming platforms are formalizing cultural consumption that was previously "invisible" to the music industry, as it existed in informal channels accessed by subaltern classes. At the same time, consumption patterns vary across countries in Latin America due to factors such as cultural preferences, socio-economic conditions, cultural policies, and market dynamics. For instance, while some countries might have a higher preference for local and regional music, others might lean toward international hits. Brazil and Mexico have emerged as key players, jointly commanding 86.4 percent of the region's music streaming revenues, with Brazil contributing with 48 percent and Mexico 38.4 percent to the total.[2]

In 2011, George Yúdice[3] noted a striking imbalance, as the Latin American music industry's share of global music revenues stood at a mere 3.8 percent, significantly below the region's then share of over 8 percent of the world population. By 2016, as streaming began to grow, Latin America's share of music streaming revenue was still just 3.6 percent. But by 2022, the region accounted for 7 percent of the global music streaming market, signaling a notable increase in influence and participation. These statistics provide a glimpse into the transformative impact of streaming on music cultures across Latin America, and how streaming offers a significant opportunity for the circulation of Latin American music cultures around the world. Moreover, Latin America's rich history of regional music consumption, shaped by a blend of cultural affinity and shared linguistic Spanish heritage, has fostered a deep connection among countries within the region, contributing to the widespread appreciation and consumption of diverse musical genres across borders within Latin America. This environment also fosters opportunities to develop national musical traditions through the emergence of blended genres.

Mexico serves as a useful case study for understanding the evolving role of MSPs in contemporary societies beyond the Euro-American mainstream global core. A distinctive feature of the country is its substantial cultural ties with the United States, primarily generated by migratory flows, especially of documented and undocumented Mexicans seeking better economic conditions. More than 50 percent of the Latinx community in the United States is of Mexican origin.[4] As a result, Mexican music and audiovisual industries generate substantial sales in the United States. This makes it important to incorporate Latinx audiences within the United States into our understanding of the circulation of Mexican cultural products, particularly music. Mexican music plays a significant role in both Latin America and the United States, acting as an important cultural pivot and bridge between both regions.

Mexico has historically served as a central hub for the circulation of Latin American music and symbolic cultural products. Mexico emerged in the early twentieth century as a central cultural node within Latin America[5] and even Ibero-America, which includes Portugal and Spain. These advantages include Mexico's distinctive geographical position and its strong economic integration with North America,[6] particularly the United States.[7] This integration has given rise to unique cultural interactions, hybridization, and migration patterns. Positioned at the territorial and imaginary intersection of the Global South and Global North, Mexico shares certain characteristics with the latter while also navigating informal organizational structures, tensions, and dynamics typical of peripheral nations.[8]

Unlike the film industry, the music sector in Mexico has historically lacked support from public authorities in the form of protectionist policies or financial assistance for production and distribution.[9] This historical discrepancy can be traced back to the influence of private interests on the Mexican political system throughout the twentieth century. Under the rule of the Partido Revolucionario Institucional (PRI) regime, which governed Mexico from 1930 to 2000, the sound and music industry remained largely under the control of nationally owned, family-based broadcasting network companies. Regulations protected national broadcasters from competition with foreign-owned companies, allowing privately owned national radio stations to flourish as a crucial window for the commercial music industry to capture audiences in the country. By contrast, the state's involvement was primarily limited to operating two public radio networks—Radio Educación and Instituto Mexicano de la Radio (IMER)—and offering minimal support for cultural expression associated with Indigenous communities.

Despite these complex dynamics, streaming music has flourished in Mexico. This chapter delves into the evolving landscape of music streaming in Mexico by addressing four key areas. The first section examines Mexico's vibrant and diverse music cultures, comparing their consumption trends in the streaming era with those in the United States and other Latin American countries. This comparison is supported by empirical research that illuminates these dynamics. The second section focuses on the unique characteristics of regional Mexican music, particularly the emergence and popularity of corridos tumbados, a contemporary subgenre that merges traditional corridos (narrative ballads) with elements of trap and hip hop. In the third section, we explore the shifts within the Mexican music industry prompted by the advent of streaming platforms and aggregators, examining changes in distribution and commercialization strategies, with a focus on corridos tumbados. The fourth section presents the perspectives of creative workers in the music industry, drawing on insights from our interviews.

1   MUSIC CULTURES AND STREAMING
CONSUMPTION IN MEXICO AND THE AMERICAS

Mexico has a very high take-up of music streaming, particularly in the vibrant metropolis of Mexico City, with its huge population of around twenty million.[10] Spotify announced in 2018 that Mexico City boasted the highest number of users worldwide.[11] Importantly, the country's consumption patterns mirror a significant global trend: the rise in streams of Spanish-language tracks. Between 2013 and 2023, there has been a remarkable 88 percent increase in Spanish-language tracks featured on Spotify's Mexico Top 100 Song Chart (fig. 6.1).[12]

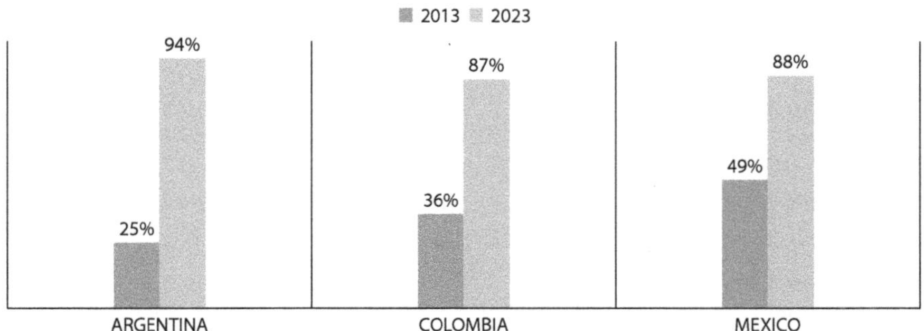

FIGURE 6.1. Percentage of tracks in Spanish in Spotify Top 100 local charts, 2013 and 2023. Generated by the authors based on Spotify data.

This consumption is primarily facilitated through free, ad-based digital platforms like YouTube or freemium services such as Spotify (i.e., platforms that offer a choice between free, ad-based use and ad-free subscription). According to ICEX data from 2023, in Mexico, Spotify is the frontrunner in the streaming music market, accounting for 30 percent of overall consumption and approximately twenty million subscribers, followed by YouTube at 19 percent.[13] This underscores the notable shift in music consumption patterns within countries like Mexico, where legal platforms have gained traction after decades dominated by informal markets. These platforms are witnessing significant growth in both ad-supported and subscription versions. Mexico also leads globally in short-format video music consumption and ranks second in terms of the proportion of online music listeners. It ranks among the top five countries with the highest percentage of users holding paid subscriptions to streaming music platforms, alongside Sweden, the United Kingdom, the United States, and Germany.[14] While most of the population—particularly those from lower socioeconomic backgrounds or those who are less engaged—accesses music primarily through free streaming platforms such as YouTube, a significant demographic segment in Mexico has sufficient purchasing power to access music via subscription. In the analog era, poorer and less engaged audiences accessed music through nonlegal channels and remained in the shadows.[15] With the advent of streaming, the consumption patterns of these groups have become much more visible.

But to what degree does Mexican music circulate via streaming platforms beyond Mexico itself, to North America and Latin America? And to what extent is music from other countries in the Americas consumed by Mexican audiences? Empirical research conducted by Alejandro Mercado-Celis, which analyzes Spotify playlists spanning three years (from January 2017 to April 2020), has provided a picture of the relationship between Mexican and US music during this period. The study explored music consumption patterns in North America (defined in this case as a socioeconomic region comprising Mexico, the United States, and

Canada), examining whether the same "successful" artists are listened to across the region. For instance, it analyzed the degree to which artists who succeeded in entering the US music charts also achieved similar success in Canada and Mexico.[16] The findings indicate that artists who successfully penetrated the music markets in the three countries represent 17.9 percent of the total North American market.[17] Furthermore, the study showed that Mexico stands out as the most diverse country in North America in terms of the national origin of popular music that it consumes.

Additionally, the study examined the nationality of artists featured in the daily lists of the two hundred most played songs in the three countries. In the Mexican case, artists with US nationality had the highest listenership, with 29 percent of the most played songs, closely followed by Mexican nationals with 25 percent. While the United States' dominance is evident, its impact during this period was less pronounced in the Mexican market compared to the Canadian one. Additionally, the research revealed that in Mexico, Spanish-speaking countries, including those from Latin America and Spain, held a dominant position in the market, representing 48 percent of the total, while English-speaking countries accounted for 38 percent.[18] Conversely, Mexican musicians accounted for less than 1 percent of the most-listened-to music on Spotify in the United States during the same period. The author remarks, "Despite the significant Mexican presence in the United States, this influence is not adequately reflected in the consumption of Mexican music within the top 200 most listened-to songs lists."[19]

However, the recent success of Latinx music has changed this picture. US Latinx music revenues surged by 16 percent in 2023, surpassing the growth of the overall market for the second consecutive year, reaching a staggering US$1.4 billion. Streaming has become the preferred choice among Latinx music consumers, with over 98 percent of the total Latinx revenues attributed to these services, according to the Recording Industry Association of America (RIAA). The same source highlighted that regional Mexican music, particularly the corridos tumbados subgenre discussed below, has been the main driver of this change.[20]

A study we conducted using Chartmetric tools casts further light on these issues. Our aim was to capture two snapshots in time and compare the performance of the top one hundred artists per country on YouTube (table 6.1) and the top fifty songs on Spotify (table 6.2) in Argentina, Colombia, and Mexico, the other major markets of Latin America. The objective was to better understand the circulation and performance of Mexican and Latin American music in global markets. We selected the weeks of March 21, 2019 (i.e., pre-COVID-19), and March 21, 2024 (post-COVID-19), for analysis. March is an ideal month due to its stability in terms of music releases, situated between two periods of intensive launches: the early months of the year and the spring releases for summer. Additionally, we included data from the United States to observe how Latin American music circulates there and to identify patterns that connect with previous research.

TABLE 6.1 Distribution of YouTube top 100 music videos by artist's country or territory of origin for major Latin American music markets
Comparison of weeks beginning Thursday, March 21, in 2019 and 2024

Artist's country or territory of origin	Mexico		Argentina		Colombia		United States	
	2019	2024	2019	2024	2019	2024	2019	2024
Argentina	2	3	29	62	2	1	0	0
Australia	0	0	0	0	0	0	1	0
Barbados	0	0	0	0	0	0	1	1
Brazil	0	1	1	1	0	0	0	0
Canada	1	1	1	1	1	0	2	3
Chile	1	2	0	4	0	0	0	0
Colombia	10	6	12	7	34	48	3	4
Costa Rica	1	0	1	0	2	0	0	0
Dominican Republic	0	0	0	0	0	0	1	0
France	1	0	2	0	0	0	1	0
Honduras	1	0	1	1	1	1	0	0
Jamaica	0	0	1	0	1	0	0	0
Mexico	48	64	7	3	10	9	3	16
Norway	0	0	0	0	0	0	1	0
Panama	0	0	0	1	1	1	0	0
Puerto Rico	12	10	23	11	28	26	6	8
South Korea	1	0	1	0	0	0	1	0
Spain	2	1	3	1	2	0	0	0
United Kingdom	3	0	3	0	2	0	4	2
United States	16	12	11	4	12	13	76	66
Venezuela	1	0	2	0	4	1	0	0
Uruguay	0	0	2	4	0	0	0	0

SOURCE: *Authors based on Chartmetric data.*

We also examined YouTube's performance, as reported by a national audiovisual consumption survey in Mexico, which found that watching music videos is the third most popular internet activity among users. The most popular MSPs in Mexico are YouTube Music (65 percent), Spotify (50 percent), Amazon Music (6 percent), Apple Music (4 percent), and Deezer (2 percent),[21] and thus, a comparison of Spotify and YouTube is crucial to understanding the broader patterns of music consumption in Mexico, among Latinxs in the United States, and in Latin America.

The data revealed a notable growth in the consumption of Mexican singers and songs in 2024 compared to 2019. This pattern of dominant consumption of national music is also evident in Argentina and Colombia, albeit to varying degrees. The consistent presence of Puerto Rican music across all three selected countries is also notable, underscoring the widespread circulation of reggaeton throughout Latin America. Colombia's case demonstrates substantial consumption of Puerto Rican

TABLE 6.2 Distribution of Spotify top 50 tracks by artist's country or territory of origin for major Latin American music markets

Comparison of weeks beginning Thursday, March 21, in 2019 and 2024

Artist's country or territory of origin	Mexico		Argentina		Colombia		United States	
	2019	2024	2019	2024	2019	2024	2019	2024
Argentina	0	0	11	41	1	2	0	0
Australia	0	0	0	0	0	0	1	0
Brazil	0	1	1	1	0	0	0	0
Canada	1	0	1	0	1	0	2	4
Chile	0	2	1	5	0	1	0	0
Cuba	0	0	0	0	0	1	0	0
Colombia	20	6	19	3	27	31	0	0
Indonesia	0	0	0	0	0	0	0	1
Italy	0	0	0	1	0	0	0	0
Mexico	5	40	3	1	1	5	0	0
Netherlands	0	1	1	1	0	1	0	0
Nigeria	0	0	0	0	0	1	0	0
Panama	0	0	0	0	2	0	0	0
Puerto Rico	19	4	23	6	27	18	1	0
Spain	0	0	0	0	0	1	0	0
United Kingdom	2	0	1	0	1	0	4	1
United States	14	14	8	2	9	7	61	47
Uruguay	0	0	2	5	0	0	0	0

NOTE: *Some songs are interpreted by two or more artists.*
SOURCE: *Authors based on Chartmetric data.*

and Mexican music. Similarly, Mexico and Argentina exhibit significant consumption of content from Puerto Rico and Colombia, in that order. At the same time, Mexico's results seem a little more diverse in terms of artists' nationalities. Equally noteworthy is the minimal presence of Anglo-Saxon music in both datasets, which affirms how local, Spanish-speaking, youth music culture scenes in Latin America are flourishing in the age of streaming.

When comparing YouTube and Spotify charts, a notable distinction arises regarding the behavior of US consumers across these platforms. Latin American musicians, including Mexican artists like Peso Pluma, are present on YouTube, particularly in the top five artists of 2024. This casts light on music consumption preferences in the United States: Latinx audiences tend to favor YouTube over Spotify. This observation aligns with Mercado-Celis's findings, reported above, which highlight the paradox of limited circulation of Mexican music on Spotify. The Mexican national preference for YouTube may offer an explanation. Generally, free Spotify and YouTube accounts dominate consumption because a significant portion of the Latinx and Latin American population lack access to bank cards.

While artists' success based on nationality is possibly cyclical,[22] it seems we are witnessing new patterns of music circulation via streaming that favor countries and artists who previously lacked such visibility and reach.

2   REGIONAL MEXICAN MUSIC AND *CORRIDOS TUMBADOS*

The emergence of corridos tumbados marks a unique development within Mexico's music culture in the age of streaming, with Peso Pluma emerging as its prominent figure. However, this genre represents the culmination of a long journey within the regional Mexican music scene. What makes corridos tumbados particularly intriguing is its adaptation by young Mexican norteños (northerners), blending traditional elements with hip hop and trap music influences in a binational context. The rise of digital platforms has given young northern Mexicans a new avenue to express and disseminate their music and cultural products. Thus, corridos tumbados are an expression of the contradictions of global capitalism's cultural flows in the age of streaming, as music crosses the boundaries of the Global South and Global North.[23]

The genre of regional music known as corridos (traditional ballads) was born during the independence movement in the nineteenth century. Its influence extends across the country from north to south, west to east. Corridos tumbados is a recent genre that blends this traditional form with elements of newer genres, such as trap and hip hop. They draw from the ballad form an emphasis on narrative lyrics, often touching on themes such as drug trafficking, love, and social issues. Reflecting the patriarchal culture of Mexican society, males predominate, often emphasizing their masculinity while also portraying sensitivity in their songs dedicated to women.

According to José M. Valenzuela, "Corridos tumbados and bélicos are recent expressions of the *corridista* tradition that have achieved enormous influence in broad youth sectors. Their narratives highlight hedonism, neoliberal consumerist codes, and drug trafficking intricacies. Social media, digital social networks, and new communication technologies are central devices for understanding the success of this music movement."[24] In the same vein, in trying to understand this new Mexican cultural music expression, Christian Fernández-Huerta understands the logics of consumption inscribed in the production and reception of corridos tumbados as devices of social distinction based on three interpretative elements: their countercultural imprint, youth realities, and the virality of digital culture.[25] The lyrics and videos of this music subgenre frequently showcase lavish lifestyles, including expensive cars, luxurious mansions, designer clothes, and high-end accessories. This display of opulence serves to reinforce the themes of success and power. Corridos tumbados blend elements of reality and fantasy to attract their audiences and listeners.

Natanael Cano, a popular singer, defines the genre as "regional Mexican like the one you, or, I don't know, your parents used to listen to; but with a younger seasoning, a younger regional Mexican [flavor]." Valenzuela defines briefly how the word *tumbado* works as follows: "Corridos tumbados recreate four meanings of the word *tumbado*: as corridos inscribed in the webs of drug delirium; as slow creations; *bajitas* as a synonym of cholo; and as productions arranged, seasoned, well done, and well *tumbadas*."[26]

Corridos tumbados' lyrics are set to a modern musical backdrop, characterized by heavy beats, electronic instrumentation, and auto-tuned vocals. This fusion of traditional Mexican folk music with contemporary urban genres reflects a dynamic appropriation by Mexican youth cultures and musicians, who integrate diverse influences and global music trends within the framework of global capitalism's contradictions, highlighting tensions inherent between the Global South and the Global North. On the one hand, there is a push for economic growth and open markets, exemplified by free trade agreements that facilitate the movement of commodities and capital across borders. On the other hand, these agreements often fail to address, and sometimes exacerbate, issues of social inequality and labor access. For instance, there are very high levels of drug consumption in the United States, while Mexico bears the burden of drug production and distribution, leading to violence and instability. Additionally, while commodities move freely across borders, people face restrictive migration policies and border controls, such as the construction of militarized border fences. The circulation of guns from the United States to Mexico further complicates the situation, fueling violence and highlighting the disparities and challenges in achieving balanced and equitable growth under global capitalism.

Corridos tumbados, though widely popular, are often criticized for glorifying drug consumption, weapons, and cartel violence—a controversy that has fueled national public debate, even as some artists now embrace themes of love and heartbreak. President Andrés Manuel López Obrador declared that the musical movement would not be censored but expressed disagreement with lyrics supposedly promoting drug use and violence: "We are not going to remain silent when they say that ecstasy pills are good, that they have a .50-caliber gun, and that their idols are the most famous narcos." The president also expressed his disapproval of an "empty, materialistic, consumerist society" that disregards the loss of young lives to drug addiction.[27]

Also important in this countercultural music expression is the role of migration flows between Mexico and the United States, along with the influence of drug cartels and narco culture, which affect youth cultures on both sides of the Mexican-US border. In other words, corridos tumbados are tied to the social and violent reality of contemporary life in the border regions. As Valenzuela notes, "For many lower-middle-class youths, the future is uncertain, and they live in the present

As noted above, the pivotal factor revolves around independent production companies associated with the regional Mexican genres. These companies typically retain autonomy in rights management, yet as previously discussed, they often forge digital distribution partnerships with major entities. Frequently, these partnerships involve artist companies associated with specific bands or artists, as exemplified by acts such as Double P, who operates under his label, Peso Pluma, or the production management handled by regional veteran Pepe Aguilar through Equinoccio Records. Additionally, independent labels within these subgenres of Mexican regional music include Del Records (home to Eslabon Armado and Lenín Ramírez), Rancho Humilde (representing Natanael Cano, renowned for corridos tumbados), Street Mob Records, and Lumbre Music.[33]

Moreover, intriguing instances of vertical disintegration emerge with the establishment of digital distribution aggregators aligned with independent labels.[34] A notable example is DSTRO7, which is affiliated with Del Records and Tamarindo Rekordsz. This initiative competes with Opplai, a music distributor and service provider headquartered in Los Angeles,[35] exclusively operating within the regional Mexican market alongside independent artists and labels such as Alianza Records and JZ Music.

Therefore, in the Mexican music industry, independent record labels and aggregators coexist with dominant major distribution companies. Independent labels tend to absorb the risk of opening new markets, discovering new genres and talent. While independent labels might self-distribute their content, they also resort to major aggregators for content distribution and tend to attract the interest of major labels as they become more successful. Independent and semi-independent players, as well as subsidiaries, serve a specific role in a differentiated relationship with better-funded major distribution companies. For instance, while major distributors are eager to capitalize on the success of the corridos tumbados subgenre, they are wary of being associated with its themes, which are often perceived as glorifying violence, and they fear that moral panics associated with the subgenre could damage companies' images and brands. This apparent contradiction is resolved by majors establishing flexible relationships with independent producers and aggregators involved with this subgenre, or by using their subsidiary aggregators to do so.

In this evolving landscape, various international players have begun to invest in the Mexican music market. The acquisition of Exile Music, a company based in Los Angeles with significant ties to the Mexican industry through artists like Vgly and Andrea Elé, by the South Korean powerhouse Hybe (renowned for managing BTS, among others), is particularly noteworthy.[36] This strategic move underscores the global significance of Mexican music and marks a pivotal leap in the transnationalization of South Korea's industry, highlighting the growing interconnectedness of music markets worldwide. A trend in which investment funds and other entities accumulate intellectual property rights through the acquisition of significant historical catalogs has also made its mark on the Mexican music scene. In 2016, the US-based company Concord, affiliated with the Michigan State Retirement Systems pension fund,[37] acquired 50 percent ownership of the esteemed Mexican label Musart.[38] Musart controls more than seventy thousand classic tracks from the annals of Mexican music, including iconic artists such as Joan Sebastian and Gloria Lasso. This acquisition reflects broader global shifts in the ownership landscape of substantial music catalogs, underscoring the evolving dynamics of intellectual property rights in the industry.

4   INDUSTRY CREATIVE WORKERS' VIEWS

We conducted nine in-depth interviews in Spanish with various professionals in the Mexican music industry, during the first half of 2024, primarily independent producers, musicians, and managers. The objective of these interviews was to gain firsthand insight into how these creative workers understand and interpret the impact of streaming platforms on the music industry in Mexico. Hearing directly from those who navigate the industry's evolving landscape on a daily basis is essential to capturing the nuanced, lived experiences that broader data or industry reports may overlook. Most of the interviewees requested to remain anonymous, which we respected to ensure open and honest reflections.

Interviewees pointed out significant transformations in two key areas: music production and distribution. Regarding the former, our informants identified a shift whereby creators or singers no longer require a traditional production house to craft high-quality recordings. Thanks to advances in music technology hardware and software, individuals can now produce their own music with relative ease. This also seems to be encouraging the mixing of music genres, apparent in the corridos tumbados subgenre we have been discussing. Streaming has also changed the temporality of music releases. Artists and bands no longer need to produce full-length albums in the same way; instead, the focus has shifted to releasing singles. Another significant change is occurring in distribution: the potential for global circulation has greatly increased, making it more efficient and contributing to the visibility of music, with social media platforms serving as a key tool to boost new artists and songs.

Regarding the growth of the Mexican industry, one of our informants highlighted the importance of "aggregators setting up their offices in Mexico for all of Latin America." Mexico has become a hub, offering proficient technical services for digital distribution and royalty collection, as well as providing digital marketing services that benefit artists and include a commission percentage for aggregators.

Another aspect that informants consistently remarked on was the emergence of binational networks or circuits connecting Mexico and the United States. One informant noted:

> For me, the genre [corridos tumbados] knows no borders between the two countries. An artist from Culiacán can have his team in Los Angeles and perform concerts in both countries. It's known that the payment per stream is higher in the United States, but generally, the largest percentage of consumption comes from Mexico. Moreover, the Mexican diaspora in the United States has brought an unmatched mix of sounds. They are generations that grew up with Anglo music like hip hop but saw their parents enjoy corridos and norteño music. This blend is no doubt just part of the exciting moment the industry is experiencing.

Additionally, they noticed that streaming platforms, such as YouTube, serve as a guide for promoters when scheduling bands, especially in circuits that heavily rely on live performances, such as regional Mexican music.

Simultaneously, while acknowledging the advantages of streaming, including ease of distribution, our informants also underscored that, in some instances, the benefits fail to translate into financial gains for composers or musicians, confirming a solid body of previous research literature from other locations.[39] For instance, the fact that a song or video on YouTube garners millions of views does not guarantee adequate economic compensation for the author or singer, especially when it comes to older songs. In this regard, the platform and the music label are the only beneficiaries of the scale of reproduction of the video or song.

Another aspect that stands out is the reported lack of professionalization in some sectors of the industry. Many actors perform several functions simultaneously, without specializing. Furthermore, interviewees emphasized the lack of support from public policies or funds, drawing a comparison with countries like Spain or Brazil, where substantial governmental backing exists for the promotion of local artists. In contrast, Mexican state support primarily targets experimental or nonindustrial projects.

CONCLUSION

Our case study emphasizes the remarkable ability of youth cultures to articulate their societal concerns and contexts through music. The rise of corridos tumbados exemplifies this, serving as a platform to express the intricacies and challenges faced by some sectors of the Mexican population, while also dynamically engaging with the experiences of Mexican migrants in the United States and the contradictions of global capitalism. Moreover, this case exemplifies how the Global North and Global South interact and overlap at different levels. MSPs have played a pivotal role in elevating the visibility of popular music, particularly within the regional Mexican genres. Unlike in the past, when some expressions relied on informal channels, MSPs have provided a legitimate avenue for showcasing popular music demand and making its consumption visible by subaltern classes that, for the industry, were previously in the shadows. Nonetheless, commercial mainstream music industry players have found avenues to capture and monetize the consumption of this genre and, to some extent, the social classes that make it possible and popular. Another aspect that stands out from our case study is that MSPs have strengthened Spanish-language music consumption in Latin America, while Latin American artists and songs are circulating with greater regional and global presence, enriching the diversity of the Latin American music scene. Finally, we note with concern that the major economic beneficiaries of music curation on a global level are the large technology companies that monetize or profit from musical consumption in various ways. These range from traditional methods, such as advertising and subscription models, to various uses of big data and data mining generated as we listen to music and create playlists.

ACKNOWLEDGMENTS

Rodrigo Gómez is grateful to the Ministerio de Ciencia, Innovación y Universidades, and the Convocatoria de la Universidad Carlos III de Madrid de Ayudas María Zambrano for the requalification of the Spanish university system for 2021–23, dated May 31, 2022, based on Royal Decree 289/2021, of April 20, 2021, which regulates the direct granting of subsidies to public universities for the requalification of the Spanish university system.

NOTES

1. David Hesmondhalgh, "Streaming's Effects on Music Culture: Old Anxieties and New Simplifications," *Cultural Sociology* 16, no. 1 (2022): 3–24.

2. International Federation of the Phonographic Industry (IFPI), *Global Music Report 2024: State of the Industry* (London: IFPI, 2024), https://ifpi-website-cms.s3.eu-west-2.amazonaws.com/IFPI _GMR_2024_State_of_the_Industry_666e61ca2c.pdf.

3. George Yúdice, "New Social and Business Models in Latin American Music," in *Consumer Culture in Latin America*, ed. John Sinclair and Anna Cristina Pertierra (New York: Palgrave Macmillan, 2012), 17–33.

4. Latinx is a neologism that refers to gender-neutral Latino (masculine) or Latina (feminine) terms, used to describe those living in the United States who were born in Latin America or have ancestry there. For Latinx studies that address issues in relation to Latin American cultural industries, see Arlene M. Dávila and Yeidy M. Rivero, eds., *Contemporary Latina/o Media: Production, Circulation, Politics* (New York: NYU Press, 2014); and Rodrigo Gómez, "Latino Television in the United States and Latin America: Addressing Networks, Dynamics, and Alliances," *International Journal of Communication* 10 (2016): 20.

5. John Sinclair and Anna Cristina Pertierra, "Understanding Consumer Culture in Latin America: An Introduction," in Sinclair and Pertierra, *Consumer Culture in Latin America*, 1–13.

6. Since 1994, Mexico, the United States, and Canada have signed free trade agreements aimed at bolstering economic ties and strengthening the North American region. The initial iteration was known as NAFTA, while the current version is referred to as the USMCA, CUSMA, or T-MEC.

7. By the end of 2023, Mexico emerged as the United States' leading commercial partner, surpassing both Canada and China. See Ana Swanson and Simon Romero, "For First Time in Two Decades, U.S. Buys More from Mexico Than China," *New York Times*, February 7, 2024, www.nytimes .com/2024/02/07/business/economy/united-states-china-mexico-trade.html.

8. In this regard, see Néstor García Canclini, *Culturas híbridas: Estrategias para entrar y salir de la modernidad* (Ciudad de México: Grijalbo, 1998).

9. Rodrigo Gómez, "The Mexican Film Industry, 2000–2018: Resurgence or Assimilation?," in Political Economy of Media Industries, ed. Randy Nichols and Gabriela Martínez (New York: Routledge, 2019), 57–82.

10. IFPI, Engaging with Music (London: IFPI, 2021), www.ifpi.org/wp-content/uploads/2021/10/IFPI-Engaging-with-Music-report.pdf; Víctor Ávila-Torres, "Making Sense of Acquiring Music in Mexico City," Networked Music Cultures: Contemporary Approaches, Emerging Issues (2016): 77–93.

11. Spotify, "Mexico City Is Now the World's Music-Streaming Mecca," For the Record, November 19, 2018, https://newsroom.spotify.com/2018-1119/mexico-city-is-now-the-worlds-music-streaming-mecca/. On the complex dynamics of Mexico City in relation to live music and its sociocultural context, we recommend Michaël Spanu, "From the 'Musical Agora' to an 'Extreme Sport': Negotiating Live Music's Values in a Context of Urban Informality, Corruption and Violence in Mexico City," Cultural Sociology 18, no. 4 (2023): 528–55.

12. As revealed by Spotify in 2023; see "Spotify's 10-Year Journey Elevating Latin Creator," For the Record, December 18, 2023, https://newsroom.spotify.com/2023-12-18/spotifys-10-year-journey-elevating-latin-creators/.

13. ICEX España Exportación e Inversiones (ICEX), *La industria de la música en México* (Madrid: ICEX, 2023), www.icex.es/content/dam/es/icex/oficinas/077/documentos/2023/11/ficha-sector-el-mercado-de-la-musica-en-mexico-2023/FS_Industria%20de%20la%20música%20en%20México%202023_REV.pdf.

14. ICEX, *La industria de la música en México*.

15. Regarding these practices or phenomena in Latin America and the Global South more broadly, George Yúdice, prior to the advent of streaming, identified them as *música paralela*. See Yúdice, *Nuevas tecnologías, música y experiencia* (Barcelona: Gedisa, 2007).

16. Unlike elsewhere in the world, the term "North America" is commonly used in Mexico to refer not only to Canada and the United States but also to Mexico, partly because Mexico is located in the Northern Hemisphere.

17. Alejandro Mercado-Celis, "Escucha digital de música popular en América del Norte," *Industrias culturales norteamericanas en la era digital*, ed. Alejandro Mercado-Celis and Santiago Battezzati (Ciudad de México: UNAM-CISAN, 2023), 101–30.

18. Mercado-Celis, "Escucha Digital," 117. It's important to acknowledge that this data may carry a certain bias stemming from the Spotify user profile in Mexico, which tends to represent urban, middle-class audiences with a notable consumption preference for US cultural products.

19. Mercado-Celis, "Escucha Digital," 118.

20. "El aumento de ingresos de la música latina en EEUU supera el global del mercado," *EFE*, April 23, 2024, https://efe.com/cultura/2024-04-23/el-aumento-de-ingresos-de-la-musica-latina-en-ee-uu-supera-el-global-del-mercado/.

21. Instituto Federal de las Telecomunicaciones (IFT), *Encuesta nacional de consumos audiovisuales* (Mexico City: IFT, 2023), https://somosaudiencias.ift.org.mx/archivos/01reportefinalencca2023_vp.pdf.

22. See Richard A. Peterson and David G. Berger, "Cycles in Symbol Production: The Case of Popular Music," *Ethnomusicology* 40, no. 2 (1975): 158–173; and Martin Kretschmer, George Michael Klimis, and Chong Ju Choi, "Increasing Returns and Social Contagion in Cultural Industries," *British Journal of Management* 10 (1999): 61–72.

23. Young northern Mexicans navigate a complex landscape marked by strong cultural identities, economic opportunities and challenges, and the pervasive influence of cross-border dynamics. Their resilience, adaptability, and entrepreneurial spirit are key traits that define their cultural identities.

24. José M. Valenzuela, *Corridos tumbados: Bélicos ya somos, bélicos morimos* (Guadalajara: NED Ediciones, 2024), 21. Unless otherwise noted, all translations are by the authors.

25. Christian Fernández-Huerta, "Epílogo: Tan pesados como una pluma: Tres claves para entender el fenómeno de los corridos tumbados," in Valenzuela, *Corridos tumbados*, 24.

26. Valenzuela, "Corrido tumbados," 26. We refer to this quote because, from a cultural studies perspective, Valenzuela explains the essence of the tumbado idea, which refers to the Mexican adaptation of hip hop aesthetics, creating a laid-back, slow mood in relation to the traditional music genre of corrido.

27. Berenice Bautista, "Natanael Cano: 'Tener más gente cantando el género nos ayuda,'" *Los Angeles Times*, July 1, 2023, www.latimes.com/espanol/entretenimiento/articulo/2023-07-01/natanael -cano-tener-mas-gente-cantando-el-genero-nos-ayuda.

28. Mireya Cuellar, "Hay que entender el entramado cultural de los corridos tumbados: José Valenzuela," *La Jornada*, January 30, 2024, www.jornada.com.mx/noticia/2024/01/30/espectaculos /hay-que-entender-el-entramado-cultural-de-los-corridos-tumbados-jose-valenzuela-9733.

29. Felipe Garrido, "The Success of Humble Corridos," Chartmetric, April 18, 2023, https://hmc .chartmetric.com/regional-mexican-music-corridos-tumbados-success/.

30. In Mexico, there are several regional music genres, such as norteño, son jarocho, mariachi, banda, and sierreña, among others.

31. Benjamin Birkinbine, Rodrigo Gómez, and Janet Wasko, eds., *Global Media Giants* (New York: Routledge, 2017).

32. Griselda Flores, "Peso Pluma's Double P Records Signs Global Distribution Agreement with the Orchard," *Billboard*, December 15, 2023, www.billboard.com/business/business-news/peso-pluma -label-double-p-distribution-the-orchard-1235559830/.

33. Griselda Flores, "Indie Labels Are Driving Regional Mexican Music's Surge: Here Are 6 That Are Leading the Pack," *Billboard*, July 31, 2023, www.billboard.com/lists/regional-mexican-music -growing-popularity-indie-labels/del-records/.

34. Allen J. Scott, "The US Recorded Music Industry: On the Relations between Organization, Location, and Creativity in the Cultural Economy," *Environment and Planning A: Economy and Space* 31, no. 11 (1999): 1965–84.

35. "La distribuidora de música latina Oplaai se expande," *Latinos Unidos*, February 10, 2022, https://latinosunidosonline.com/la-distribuidora-de-musica-latina-oplaai-expande-su-portafolio -para-incluir-servicios-de-management-y-disquera/.

36. Sohee Kim and Lucas Shaw, "K-Pop Giant Hybe Buys Its First Latin Music Company," *Bloomberg*, November 13, 2023, www.bloomberg.com/news/articles/2023-11-12/k-pop-giant-hybe-buys-its -first-latin-music-company.

37. Brian Croce, "Michigan Retirement Likes the Sound of Its Concord Music Stake," *Pensions and Investments*, April 29, 2019, www.pionline.com/article/20190429/PRINT/190429867/michigan -retirement-likes-the-sound-of-its-concord-music-stake.

38. Ed Christian, "Concord Bicycle Music Acquires Mexican Indie Musart Music Group," *Billboard*, April 21, 2016, www.billboard.com/pro/concord-bicycle-music-acquires-musart-music-group/.

39. See David Hesmondhalgh, "Is Music Streaming Bad for Musicians? Problems of Evidence and Argument," *New Media and Society* 23, no. 12 (2020): 3593–3615. Another interesting example is Andrew deWaard, Brian Fauteux, and Brianne Selman, "Independent Canadian Music in the Streaming Age: The Sound from Above (Critical Political Economy) and Below (Ethnography of Musicians)," *Popular Music and Society* 45, no. 3 (2022): 251–78.

7

The Japanese Transition to Streaming Music

Corporate Hesitancy and Individual Innovation

Noriko Manabe

While the global music industry has undergone a significant transformation to streaming over the past two decades, Japan has notably lagged in the adoption of streaming services. According to the International Federation of the Phonographic Industry (IFPI), streaming accounted for 67 percent of global music revenues in 2023, while physical sales have dwindled to 18 percent; in Japan, these figures are flipped, with physical sales still accounting for 65 percent of revenues and streaming accounting for 31 percent—roughly equivalent to the US market eight years prior, in 2015 (fig. 7.1). Japan's music market remains heavily reliant on physical media, whose sales were up a robust 9 percent year over year in 2023.

In 2016, Ono Tetsutarō, who later became the CEO of Japanese streaming service Awa, told me that it might take five to eight years for the streaming market in Japan to reach maturity.[1] Eight years later, music streaming in Japan appeared to be approaching maturity, with revenue growth decelerating to 14 percent year over year in 2023, but the penetration of paid streaming services remained significantly lower than global averages. According to a 2023 survey by the Recording Industry Association of Japan (RIAJ), only 26 percent of Japanese consumers used paid streaming services, compared to 48 percent globally.[2]

This chapter examines the factors contributing to the relatively slow growth of music streaming in Japan, exploring both supply-side issues within the music industry and demand-side factors among Japanese consumers. As the second-largest music market in the world, Japan presents a valuable case study on how the proliferation of new media technologies is not a given but is instead shaped by the interaction of corporate and cultural forces. Corporate strategies and consumer lifestyles have shaped the way media is conceptualized in Japan, which has

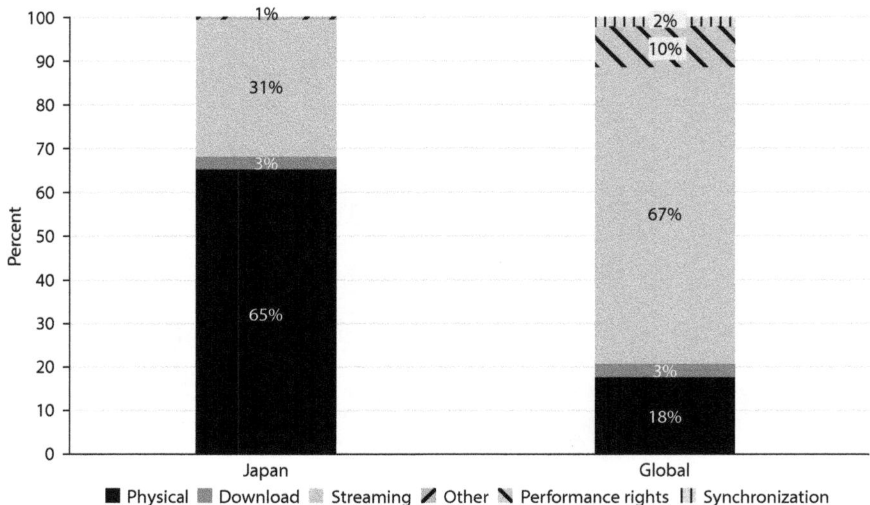

FIGURE 7.1. Japanese versus global music industry revenues, 2023. Generated by the author based on data made publicly available by the Recording Industry Association of Japan (RIAJ) and the International Federation of the Phonographic Industry (IFPI).

impacted the acceptance of streaming. Corporations and artists have resisted the paradigm shift from manufacturers selling products (new music) to rentiers profiting from owned property, which reflects the capitalist logic of music streaming.[3] At the same time, independent artists have capitalized on internet platforms to reach a global fan base that was inaccessible to them in the Japanese music industry ecosphere.

REASONS FOR THE SLOWER GROWTH OF STREAMING RELATIVE TO GLOBAL LEVELS

Japan's lower adoption of music streaming can be attributed to factors on both the supply side (within the music industry) and the demand side (among consumers).

Supply-Side Factors

The shift from an ownership model to an access model of music distribution represents a fundamental change in the music industry's economic paradigm. As Eric Drott argues, this transition has profound implications for how value is created and extracted in the music economy.[4] In the traditional CD-based model, music is treated as a discrete product to be sold and owned. In contrast, the streaming model treats music as a service to which listeners rent access. This shift challenges the long-established business practices of the Japanese music industry.

Focus on Physical Sales

For decades, the industry has built its strategies around the sale of physical products, particularly CDs, which have remained very profitable. Since 1953, Japan has applied a resale price maintenance system to media, including recorded music, which legally requires retailers to sell CDs at the list price without discounts. As CD prices are relatively high (approximately ¥3,000 for albums and ¥1,000 for singles), the system ensured profitability for record companies. Japanese record companies developed a highly effective system for creating and promoting hit songs through "tie-ups," where new songs are often introduced as theme songs for television programs or commercials, which are then repeatedly aired. In a country where cable television never caught on to the same degree as in the United States, terrestrial television was an overwhelming influence on cultural trends and remained a strong medium through the 2010s; as late as 2018, two-thirds of Japanese were listening to music through television.[5] When a song became well-known, record companies would advertise it on TV. Their aim was to boost the album or single to a high rank on the Oricon charts, Japan's most-recognized sales ranking. Achieving such a high rank would prompt television shows and magazines to feature the artist, thereby further boosting sales. CD sales were supported by exclusive dealership contracts between record companies and CD specialty stores, which allowed for exchange of information, careful inventory management, and coordinated shipments according to sales data. Even in 2024, CD and vinyl retailers remain a fixture in Japanese cities, with multistory brick-and-mortar locations for Tower Records and HMV and many used record shops dotting the landscape.

Given the power of the Oricon chart, the music industry conducted business in a way that would maximize this ranking. Despite a robust market for mastertones in Japan since the mid-aughts, Oricon famously resisted including downloads or streaming in its charts until December 19, 2018; until then, its charts only counted physical sales. Hence, record companies were not incentivized to grow internet-related sales and focused instead on maximizing CD sales. By the 2010s, there was little correlation between the Oricon charts and what people were listening to; the chart was rewarding the marketing of CDs, not listening habits.

One of the most common marketing ploys used by Japanese record companies is placing premiums inside CDs. These premiums are similar to the toys placed in cereal boxes before the 1990s, where the toy was the premium that made the cereal desirable. For example, Japanese record companies often release multiple versions of the same album with different bonus tracks, attach a bonus DVD, or include application forms for concert tickets with CD purchases; the desire for these premiums would induce some dedicated fans to purchase multiple copies of the same CD.

This practice was executed most successfully with AKB48, a hundred-member idol-pop group, which was founded on the concept of "idols you can meet." Certain CDs contained tickets to meet-and-greets that would allow the holder ten seconds with the AKB48 member of their choice; forty-year-old men would buy ten copies of the same CD so that they could have more time with their favorite girl. This experience of meeting the idol was the premium—not the song, which was just the ticket for the experience. AKB48 went a step further by having a general election by fans that decided which girls would be featured as lead singers or soloists. As the ballot was contained in a particular CD single, fans would buy multiple copies of the same CD to vote up their favorite member. In 2014, one forty-two-year-old farmer bought 4,600 copies of the same CD for the election.[6] For much of the 2010s, AKB48 and its sister groups dominated the Oricon charts; for the yearlong chart in 2013, they took seven out of the top ten slots. Hence, the Japanese music industry geared its systems to sell as much physical product as possible, which would maximize profits. For this reason, CD sales in Japan never declined to the low levels seen elsewhere. Yet the charts did not reflect actual listening habits.

Hesitancy to Supply Content

The high profitability of physical media has historically made Japanese music companies hesitant to support internet platforms. Indeed, one of the most significant barriers to streaming growth in Japan has been the reluctance of many record labels and rights holders to make their catalogs available on streaming platforms. The Japanese music industry has often viewed the internet with suspicion, regarding it as a cauldron of piracy—a problem in Japan, though it has never reached the extent seen in the United States or elsewhere.[7] It has been more concerned with the low unit prices of internet business models and cannibalizing physical sales. The industry has also been wary of giving up the more direct relationship with listeners fostered by its ecosystems of CD sales, close retail partnerships, and fan clubs; instead, the internet platform would serve as the central hub for information. This hesitancy has resulted in notable gaps in the catalogs of internet-based services, particularly for well-known Japanese artists, which has slowed the adoption of these platforms.

This reluctance to supply content also impacted the penetration of iTunes Japan, whose growth was significantly slower than in other countries. Opened in August 2005, iTunes was expected by Japanese record companies to result in lower revenues, as unit prices were lower and users could download a single track rather than the entire album. To obtain product for the iTunes music store, Apple had to secure master recording rights, for which there is no organization that centrally manages and grants them; these rights were typically owned by the artist agency or record company, which could refuse them.[8] As a result, the catalog on iTunes Japan was incomplete compared with iTunes in the United States. Several major record companies, including BMG Japan and Warner Japan, were not available when it launched, and Sony Music Entertainment Japan, which operated the competing download site Mora, withheld its titles from iTunes Japan for seven years, only relenting in November 2012. Back catalogs also took many years to be more widely available. Some artists, like the rock group Southern All Stars, one of the most popular and influential bands of the postwar era, withheld their songs from iTunes Japan and Mora until December 2014. Users thus found it difficult to find songs by their favorite artists on iTunes Japan. Furthermore, iTunes Japan was not price-competitive compared with CD rental shops, where one could rent (and copy) an entire album for ¥280, while individual tracks on iTunes cost ¥150 at the time of its launch. These conditions hampered the growth of iTunes in Japan.

Similarly, record companies were reluctant to provide content to streaming services. In the United States, online radio services such as Pandora (which do not play specific songs on demand) were able to work under a statutory license without securing explicit permission from copyright holders, giving them rapid access to a large catalog. This license allowed online radio to permeate, laying the groundwork for on-demand services like Spotify. In Japan, no such statutory license existed, so companies had to secure permissions for recordings from the artist agency or record company. Pandora was never able to gain a foothold in Japan, and online radio, like streaming music, never really caught on, despite several attempts from the late 2000s onward.

Difficulties Confronting Streaming Services

The year 2015 marked a turning point for streaming in Japan, as several major local services launched. AWA debuted in May 2015 as a joint venture between the Japanese music and entertainment company Avex and the digital media company CyberAgent, followed by Line Music in June 2015, backed by the messaging platform company Line Corporation, Sony Music, and Universal Music. These local services sought to establish a foothold before the anticipated arrival of Spotify, which the media referred to as the "Black Ship"—a reference to US Commodore Matthew Perry, whose landing in 1853 forced Japan to open to trade. AWA and Line Music were quickly followed by Japanese subsidiaries of US companies, with Apple Music entering the Japanese market in July and Amazon launching a music service tied to Amazon Prime in November 2015. Notably, Spotify had incorporated its Japanese subsidiary in 2013 but didn't launch until September 2016, delayed by protracted licensing negotiations with Japanese labels.

In 2016, Ono of AWA seemed cautiously optimistic about prospects for streaming music. The labels recognized that Spotify and Apple Music would inevitably launch in Japan, and pressure from foreign competition pushed them to make more of their catalogs available to streaming services, even for some popular songs.

Nonetheless, the availability of Japanese content remained the key impediment, repeating the experience with iTunes Japan. By 2016, most major streaming services offered global catalogs of over forty million tracks. However, Japanese content remained more limited: out of AWA's thirty million tracks in 2016, only about five hundred thousand were Japanese songs, primarily consisting of older catalog material from the 1960s onward. This balance was problematic in a music market where 89 percent of production value in Japan came from domestic rather than international artists.[9] A mid-2018 *Nikkei Trendy* survey of Amazon Music Unlimited, Apple Music, Line Music, and Spotify found that while these services offered over forty million titles, their libraries still did not provide comprehensive Japanese music catalogs.[10] It compared the availability of thirteen popular Japanese artists, pulled from rankings of favorite artists on the Oricon and Recochoku charts for 2017, across these different services. Only about half the artists were available on each platform, and several artists were not available on any—the boy-idol agency Johnny and Associates' artists Arashi and SMAP, J-rock bands B'z and Back Number, and J-pop artists Hoshino Gen and Amuro Namie. Indeed, several top-selling agencies had held out from streaming services. Johnny and Associates famously withheld any product from the internet, including iTunes, until founder Kitagawa Johnny died in 2019. As of 2024, several artists managed by Hello Project, such as the popular idol group Morning Musume, remained unavailable on Spotify. These challenges in providing a complete catalog posed a problem for the streaming companies: in a world where young people watched YouTube rather than terrestrial television, musical taste had become fragmented, obligating a streaming service to offer a widely varied catalog: "Customers get mad when their favorite tracks or artists are not on the service."[11]

In Japan, artists and agencies discuss with record companies whether to provide music for streaming, but the companies sometimes postpone the decision and reduce the number of songs made available. Streaming embargoes were often at the individual artist's discretion. Some feared that their CDs would not sell; some older artists valued the personal interaction of receiving money from a fan and handing them a CD, feeling that a simple data transmission devalued the music. The Southern All Stars did not make its songs available for subscription-based streaming until December 2019, and, as mentioned above, the stable of Johnny and Associates kept its materials off the internet until that year as well. This initial hesitancy, however, wasn't limited to older, established artists. Even some younger acts withheld their music from Japanese streaming services while making it available on global platforms. In 2016, the popular rock band One OK Rock was allowing its music to be streamed internationally on services like Spotify but not in Japan. The band used streaming for promotion in overseas markets where it was less established, while protecting CD sales in Japan, where it saw streaming as financially risky.[12]

Reluctance toward the Freemium Model

Many internet businesses have used the freemium model, offering a free tier with advertisements and limited features to attract new users, develop their familiarity with the service, and convert them to paid, premium-tier users. Spotify notes in its annual reports that freemium users are the primary source for premium subscribers.[13] The Japanese streaming market has been hesitant to follow suit.

Ono, of AWA, explained in 2016 that while he personally saw potential in the freemium approach, Japanese record labels were strongly opposed to the idea not only because it devalued music; it also formed a three-way relationship between the label, the listener, and the advertising company, where the advertising company paid so that the listener could access music for free. The labels preferred maintaining a direct relationship with the listener, who paid to listen to music.

However, Ono seemed to be skeptical about the economics of the freemium model, noting that Spotify was in the red because of its free tier. Indeed, even as late as 2023, Spotify was losing money on its free tier: it was generating just enough ad-supported revenues to cover ad-supported royalty payments, leaving a large loss after allocating administrative and R & D costs.[14] While Ono recognized that freemium strategies expanded the market by taking such losses upfront, he wondered how long investors and management could tolerate red ink.

Consequently, most Japanese streaming services avoided freemium offerings in their initial launches, instead offering limited free trials. For example, AWA's free tier initially allowed only one hour of free listening per month (without on-demand capabilities and with advertisements) before extending the limit to twenty hours per month in late 2016, to remain competitive with Spotify Japan. When Apple Music and Google's streaming service launched in Japan, they did so without free tiers, in line with this market preference, while Amazon had a lower-cost version (with a much smaller catalog) for Amazon Prime members. When Spotify finally launched in Japan in late 2016, it became the first streaming service to offer a free tier with unlimited listening time.

Free tiers that were later introduced seemed more restrictive than what one might expect in North America or Europe. AWA's late-2016 configuration (still in place in 2024) offered on-demand capability and no ads but limited the listener to ninety-second highlights per song. Line Music, a subsidiary of the popular messaging app Line, briefly experimented with a limited free tier from 2020 to 2021, but as of 2024, it only offered thirty-second previews to nonsubscribers. As of 2024, Amazon Japan offered a free tier with unlimited listening hours but with limited on-demand capabilities, which was similar to Spotify Japan's stance as of 2024. The relative unavailability of a freemium model may have limited consumers' ability to get to know a service, which would have enhanced overall diffusion.

Competition among Streaming Services

It is difficult to obtain verifiable market shares for streaming services in Japan, as none of them breaks out its streaming revenues in Japan in financial statements, and user surveys produce different results, depending on the surveying company.[15] Nonetheless, most surveys show that the top three streaming services in Japan are Spotify, Apple Music, and Amazon (specifically Amazon Prime), with YouTube Music, Line Music, AWA, and other services further behind. While the freemium model appears to have helped Spotify pull in customers and gain user share, the service does not have the dominant share in Japan that it enjoys in Europe or the United States. A user survey in Japan by the digital marketing company Nyle suggests that Spotify had perhaps 23 percent of user accounts in 2023; per Midia Research, its share of the global subscriptions market was 31.7 percent in the third quarter of 2023.[16]

In the Japanese market, Apple and Amazon have benefited from the name recognition of their global brands. Other services have sought to differentiate themselves through features: Line Music, which is popular among tweens and teens, leverages the ubiquity of its messaging app to offer background music that plays on one's profile page.

But in 2024, even as freemium models in Japan had become more widely accessible and most artists (except for a few holdouts) had made their products available, the diffusion of streaming services in Japan remained relatively low. The Nyle survey found that 47 percent of Japanese between the ages of fifteen and sixty-nine used streaming services; this paled in comparison to the IFPI's finding that 73 percent of global listeners used licensed streaming services.[17] What about Japanese consumers' behavior was slowing the growth of streaming?

Demand-Side Factors: Consumers

Several demand-side factors—rooted in demographics, cultural practices, and consumer preferences—explained the slower growth of streaming services in Japan compared to other major music markets.

Demographics

Japan's demographic profile presented a significant challenge for the expansion of streaming services. With almost 30 percent of its people aged sixty-five or older in 2024, Japan had one of the oldest populations among major economies; this compared with 18 percent in the United States.[18] In contrast, those under the age of twenty-five have historically been the most avid music listeners (and buyers) as well as adopters of new technologies; the share of people aged ten to twenty-four in Japan was only 13 percent, compared with 19 percent in the United States. Indeed, the IFPI user survey showed that while over 60 percent of those between sixteen and thirty-four globally had streaming subscriptions, this figure fell to 28 percent for those between fifty-five and sixty-four. Similarly, while 32–34 percent of Japanese aged twelve to twenty-nine paid for streaming subscriptions, only 14 percent of those over sixty did.[19] Japan's demographics suggested that its market was less inclined to adopt new technologies or alter long-established music consumption habits compared to countries with younger populations.

Culture of Engaged Listening

In his book *Streaming Music, Streaming Capital*, Eric Drott notes that the metaphor "streaming music" likens music to water, betraying an ideology that devalues it as a kind of utility—a ubiquitous background that is not fully appreciated.[20] But this is counter to how many Japanese experience music—in a more engaged mode of listening. While many surveys show that Japanese people often listen to music while performing other activities (e.g., commuting, exercising, doing housework), there is also a strong tradition of focused, attentive listening that differs from the passive background listening common in other markets. Indeed, one executive from Mixi Music, an online radio service affiliated with the Mixi social network, attributed its failure in 2009 to its nonsocial, noninteractive nature: "Applications on Mixi tend to be most successful when they reinforce the connections between people. If you are streaming music from your PC in Japan, you are probably listening to it by yourself. It wasn't communicative, like participating in an activity with your friends."[21]

Japan's culture of music consumption values engaged, attentive listening. One example is its long history of *kissaten* and listening bars—cafés or bars where a knowledgeable proprietor has a voluminous collection of records specializing in a particular genre and a superior sound system, and the patrons gather to listen to carefully curated selections of music. Some have rules against talking. Such *kissaten* have been instrumental in introducing genres like jazz, rock, and reggae to Japan: they enabled musicians to learn the music—some even transcribed it as they heard it—and form connections with those who became band members, producers, and entrepreneurs in the genre. Since the 2010s, when the police began cracking down on dance clubs, small listening bars featuring electronic dance music with DJs have popped up as places for people to gather and listen.

Several Japanese DJs working in Europe have mentioned to me that Japanese clubs have a culture of close listening that European audiences lack. In Japanese clubs, many audience members face the stage, and a group of dedicated fans encircle the DJ, watching their every move. The superior equipment in Japanese clubs, relative to European clubs, allows DJs to cater to this closer-listening culture with more subtle mixes. In Europe, audiences are more interested in socializing and tend to face one another in a circle rather than facing the DJ.

Furthermore, Japan does not have the kind of radio culture that would have developed a culture of passive listening. In car-focused cultures like the United States, people often listen to the radio on their way to work or while running errands. Across the globe, 76 percent of the population listens to music on the radio, whereas in Japan, only 16 percent of music listeners do.[22] Most radio in Japan is structured like television, with programs in numerous time slots, each presenting a different genre. Moreover, the programs often focus on personalities and resemble a talk show featuring music. They simulate having a guest in one's living room who introduces music, rather than serving as a passive source of music. The attention is on the discussion. The differences in radio culture may have affected the adoption of streaming services, as the concept of "online radio" as a hostless, algorithmic playlist had less cultural resonance. Taken together, these listening habits may make Japanese consumers less inclined to view music as background entertainment, thereby potentially reducing the appeal of streaming services.

Attraction of Physical Ownership

When brick-and-mortar music stores disappeared around the globe in the 2000s, Japan remained an anomaly, with not only Tower Records but also local record shops dotting the landscape; in these record stores, I would encounter American DJs like Pete Rock, who made a point of digging whenever he was in Japan. Indeed, Japanese consumers, particularly older ones, have shown a strong attachment to physical media. The foremost reason they give for purchasing physical media is to support their favorite artist and to add to a collection; many also want the premiums included with CD purchases.[23] Hence, purchasing physical media remains an aspect of fan engagement.

Some older customers profess discomfort at paying so little for music. Suzuki Osamu, an entertainment businessman in his late forties, mused in a 2019 article that he liked the convenience of instant access afforded by streaming services but also felt guilty about using them: "For those of us who bought records and CDs in the '80s or dubbed what we rented from rental stores, listening to music for free breaks our hearts. Every month, you can listen to as much music as you want for the price of a single CD."[24]

Music as Experience

Lastly, music is often presented as part of a larger sensory experience in Japan. Patrick St. Michel notes how CD stores have transformed into hybrid spaces that offer concerts, cafés, and other amenities.[25] Tsutaya's stores in fashionable Tokyo neighborhoods like Daikanyama and Roppongi are not just music and book stores; they offer a sophisticated, style-conscious experience, complete with cafés and plush lounging areas. On my visits to these stores, I have seen many couples on dates. Large stores like Tower Records in Shibuya have ample space for concerts, events, and livestreams, as well as cafés and bars with music-themed drinks. These retailers underline not only the persistence of physical media in Japan but also the fact that music consumption is approached as a special occasion rather than a background activity. These factors contribute to a consumer environment in which the value proposition of streaming services does not always align with established listening habits, as it has in other markets.

IMPACT OF INTERNET/STREAMING

Despite the slower adoption of streaming, the advent of such platforms has had a notable impact on consumption habits, artist promotion strategies, and industry practices in Japan.

Changes in Consumption Habits

Despite the resistance, 2016 still marked an inflection point in Japan's transition to streaming. Nestled among the predictable entries by AKB48 and Johnny's idol groups on the Billboard Japan chart for that year was "PPAP"—a short novelty song by Pikotaro that went viral on YouTube. Indeed, by the 2010s, YouTube had become established as a dominant platform for music listening in Japan, with 59 percent of respondents in the 2023 RIAJ survey reported having used it in the past six months.[26] Its growth broke the stranglehold that terrestrial television had on pushing new artists and songs.

Adding to this diversification of taste were streaming services. At an early point in the diffusion cycle, AWA found in its 2016 survey that 83 percent of its users claimed to be listening to music for longer periods after adopting the service, as well as listening to a wider range of artists—an average of about ninety artists per month. Ono pointed to these statistics as a sign that streaming services were fostering a more exploratory listening culture, leading to the discovery of new artists.[27]

The COVID-19 pandemic accelerated changes in music consumption habits in Japan. Traditionally, artists had released CDs and promoted them through appearances on TV, live performances, and events at record retailers. The pandemic led to the cancellation of such appearances, as well as the postponement of CD releases. The focus shifted toward digital content and online performances, creating opportunities for artists to gain visibility through internet buzz. This period saw the YouTube channel The First Take, in which artists record a performance live on one take, gain significant popularity. A video of the duo Yoasobi performing "Racing into the Night" ("夜に駆ける"), filmed at home and posted in May 2020, shot to the top of the Billboard Japan streaming charts and made Yoasobi stars without their having ever released a CD.

The rise of short-form video platforms like TikTok has created new pathways for Japanese artists to gain popularity, bypassing traditional industry gatekeepers. In 2020, the indie artist Eito self-released "Perfume" ("香水"), which went viral on TikTok, reaching number one on the Oricon chart, and Eito was given a coveted appearance on NHK's New Year's Eve Red-and-White Song Contest. Interestingly, the song attracted attention more than six months after its release, reaching fame when fans used it to accompany their own TikTok posts. Some artists have even created songs specifically with TikTok virality in mind.

Streaming has also led to a greater appreciation of older music. According to the Luminate survey, the share of album consumption for catalog (releases more than eighteen months old) in the United States was 72.6 percent in 2023; catalog album consumption rose 13.2 percent year-over-year in unit terms, while current music rose at a slower pace, at 10.9 percent year over year.[28] Part of this is because the internet and playlists afford ease of discovery, but it is also due to the fact that streaming-oriented charts are measuring actual listening, as opposed to sales-oriented charts, which measure who is most successful at selling CDs. The top-streamed songs are often those that have been out for some time, as people listen to a song repeatedly over time.[29]

In a streaming environment, songs suddenly become viral months, years, or decades after the recording, which reorients the concept of a release. The IFPI notes that 64 percent of listeners between the ages of sixteen and forty-four like discovering older music.[30] With discovery through playlists and social media rather than television programs, it matters less when the work was recorded; from the user's point of view, the release date is the moment they discover the song.

A dramatic example of the internet's impact on catalog is Japanese city pop, a disco- and R & B–inflected style from the 1970s and '80s. Long known among Japanese and European record collectors, the genre resurged to global prominence when an unauthorized YouTube upload of Takeuchi Mariya's "Plastic Love" (1984) became a viral hit in 2017, garnering over twenty-four million views; in 2020, Matsubara Miki's "Mayonaka no Door (Stay with Me)" (1979) reached number one on the Spotify global viral chart.

International Visibility

The global city pop phenomenon suggested that the appeal of Japanese music was broader than previously assumed by the domestic industry. For much of the 2000s and 2010s, the Japanese music industry was uninterested in overseas markets. There were high-profile flops by top Japanese artists who released work in the United States (e.g., Pink Lady, Matsuda Seiko, and Utada Hikaru). More importantly, the labels were making so much money in the Japanese market that they saw little attraction in other markets, which were either small relative to the Japanese market (particularly in many Asian markets) or were seen as bastions of piracy, making them seem more risky than profitable. The city pop boom demonstrated that it was time for the Japanese industry to put these concerns aside and explore the potential for Japanese music to find international audiences through digital platforms.

The Japanese industry's international reach is still small: for independent artists served by the music distributor TuneCore Japan, overseas revenues only accounted for 13 percent of the total. But overseas revenues are growing: the share of Japanese-language music in the global top ten thousand streaming tracks list increased from 1.3 percent in 2022 to 2.1 percent in 2023.[31] Indeed, Japanese music has global appeal, driven by fandom of video games, anime, and Vocaloids (avatars of voice synthesizers) like Hatsune Miku.

Yoasobi exemplifies the increasing global popularity of Japanese music through streaming platforms. As mentioned above, the pop duo's debut single "Racing into the Night" became a massive international hit in 2020, helped substantially by TikTok memes and the release of an English-language version. Their song "Idol" (2023), the opening theme for the anime series *Oshi no ko*, was the most-streamed Japanese song in 2023; in 2023–24, they toured Asia and the United States, playing at the Coachella and Lollapalooza music festivals.

Responses of the Record Industry

So far, the responses of the Japanese record industry betray a continued orientation toward maximizing product sales rather than rental payments. An Avex executive told me in 2016 that the company would likely expand its focus on the "upstream" aspects of the music business, particularly intellectual property and rights management. He thought it would concentrate on owning and managing music copyrights, partnering with local companies for distribution and promotion, rather than handling these aspects directly. This approach could involve managing artists internationally without necessarily being involved in all aspects of production and distribution and leveraging existing relationships with international songwriters and producers to create revenue streams from global hits.

But a look at Avex's strategic plan for 2022–27 shows a continued concentration on developing new talent and acquiring properties rather than maximizing return on the artists and catalog it already has.[32] The presentation does not enunciate a strategy for marketing its catalog or existing artists overseas; the strategy seems to be more about diversifying by developing anime, forming new boy bands specifically to appeal internationally, establishing US labels focused on American artists, and acquiring publishing rights for young US producers. This acquisition-and-diversification strategy contrasts with that of the Warner Music Group, which is more focused on the music business itself and maximizing return on the existing catalog.[33]

Indeed, record companies initially responded to the launch of streaming services in 2015–16 by treating them as another way to promote CD sales. For example, Avex first released Hamasaki Ayumi's album *Made in Japan* on streaming platforms, then held live performances before the CD release. This strategy allowed fans to learn the songs before concert performances, leading to more enthusiastic audience reactions and stronger CD preorders. Avex and other labels also began following this approach. Record companies also released individual songs rather than whole albums in order to build awareness—a shift away from the album-centered approach that had traditionally dominated the Japanese music industry.

But perhaps the biggest sign of the slow transition to a rentier mindset was the Japanese music industry's relative reluctance to grant distribution rights for its catalog. During the city pop boom of the late 2010s, multiple Westerners making compilation albums of Japanese 1970s and '80s city pop were surprised to find that some record companies and artists were refusing to make their tracks available. As of mid-2024, one of the global favorites of city pop—Yamashita Tatsuro—appears not to have any intention of making his catalog available on streaming services, saying that the people who profit from streaming are not the musicians.[34] He has achieved global popularity through the plethora of unlicensed uploads on YouTube.

Another sign that the CD mindset has not ended is royalty rates. For CDs, standard royalties in Japan are 12–16 percent of sales for the master rights and 6 percent for the copyright (to the publisher, lyricist, and composer); the performer typically receives about 1 percent of sales (taken from master rights). For streaming services, copyright royalties are 12 percent and master royalties are around

55 percent of sales; however, as streaming is often not explicitly addressed in older contracts, most labels apply the low royalty rate of CDs to streaming, at about 1 percent of revenues.[35] This figure is much lower than performers' royalties for downloads (as on iTunes), at 1.5–4 percent of sales; it is also not reflective of costs to the label, which need not carry inventory for streaming revenues.[36] It is also low relative to Europe. According to a report published in July 2021 by the House of Commons, master holders in the United Kingdom get 55 percent of revenues, as in Japan, but the performer normally gets about 16.5 percent, versus 1 percent in Japan; a UK songwriter would get an additional 10.5 percent.[37]

Many artists and managers have called for more equitable revenue-sharing in the streaming era, arguing that the "third-party usage" clause in recording contracts—which grants artists 10–20 percent of the revenue when their music is licensed for use in TV, film, and commercials—should also be applied to streaming. However, labels often insist on applying the lower CD royalty rate instead. There are growing calls to introduce unwaivable remuneration rights for streaming, following developments in Europe, to ensure fair compensation for artists.

Such situations are kept in place because of power imbalances. The Japanese music industry has tended to behave in lockstep. Although the RIAJ lists eighteen major labels in its membership, in practice, many of the smaller majors follow Sony's lead. Several others are parts of larger media conglomerates—for example, Pony Canyon is part of Fuji Sankei, which includes a television network and newspaper. The power, connections, and information such companies hold far outweigh those of individual artists.

The Rise of Independent Labels and Artists

The digital era has created new opportunities for independent labels and artists in Japan. Digital platforms allow independent artists to reach audiences without the backing of major labels. Japanese independent labels, historically overshadowed by major labels and agencies, are beginning to shine in the digital and global era.

Streaming platforms have made it easier for artists in niche genres to find and cultivate an audience. The "long tail" economics of streaming mean that even highly specialized music can find its audience and potentially generate meaningful revenue. Social media and platforms like Bandcamp allow artists to build direct relationships with fans, reducing their dependence on traditional industry intermediaries. Streaming services have also enabled independent artists to easily reach international listeners.

Indeed, for Spotify, the global share of streams from DIY aggregators and indie labels with direct Spotify deals rose to 26 percent in 2023, double what it had been in 2017.[38] TuneCore Japan, which handles internet distribution, publishing, promotion, and rights management for many indie artists, saw its share of Japanese digital music revenues rise from 9.1 percent in 2020 to 13.3 percent in 2023; it is the third-largest company in Japanese streaming, ahead of several major labels.[39]

The ability to directly access listeners has made independent artists question the value of signing to a label. The Vocaloid producer Kikuo noted that, as an independent artist rather than one signed to a label, he not only earned more money but also had the freedom to make the music he wanted.[40] Indeed, record companies hold the right to grant permission for digital distribution, but they may make conservative decisions that risk opportunity losses for artists.[41]

The case of Kikuo illustrates how the internet and streaming platforms have created new pathways to success for independent artists in Japan and beyond. His song "Love Me, Love Me, Love Me" ("愛して愛して愛して," 2013) was the first Vocaloid song to pass the hundred-million-streams mark on Spotify. As of March 2025, he had 1.7 million monthly listeners on Spotify and 1.4 million subscribers on YouTube. Kikuo got his start in desktop production as a teenager, uploading tracks to the 2chan bulletin board. In contrast to the typical cheerfulness of J-pop, his Vocaloid songs are dark, addressing suicide, death, bullying, and toxic parents. "Dance of the Corpses" (2013) invites "people who want to die" to "leave this painful world" for a "sweet paradise." "You're a Worthless Child" is about a child, bullied by his toxic mother, who "leaves"—possibly through suicide. This content would normally cause controversy with the Recording Industry Ethics Regulatory Commission, which inspects all recordings by RIAJ members before release.[42] In addition, Kikuo's social awkwardness would make it difficult for a label to promote him in the traditional way of relentless contact with fans. Nonetheless, the uniqueness of Kikuo's sound—a blend of unusual instruments, chromatic progressions, catchy melodies, and microtonal experimentation—has captured an international audience. Despite his songs being solely in Japanese, 75 percent of his listeners are from outside Japan.[43] He has toured five continents and played American festivals like South by Southwest.

Kikuo does not fit the traditional mold of the Japanese music industry, which still aims to sell a tangible product, but he understands the rentier paradigm of today's streaming music market and is better positioned to capitalize on this system. By creating unique tracks, retaining their ownership, and directly reaching a global audience, he has been able to build a sustainable career outside of the label system.

CONCLUSION

The Japanese music industry has transitioned more slowly to digital formats than other countries, due to the extraordinary profitability of the CD business model and the industry's reluctance to make its content available. Nonetheless, this transition has occurred, pushed by competition from global entrants and the COVID-19 pandemic. The internet has created opportunities for Japanese artists and older catalogues to reach a global audience. Japan's aging demographics may limit the upside on new technologies and consumption models, leaving the potential growth to overseas markets. The achievements of artists like Yoasobi and Kikuo illustrate how digital platforms can be leveraged for both domestic and international success without the need to release CDs, thereby bypassing traditional industry gatekeepers.

Their approaches suggest that the future of the Japanese music industry may involve a reconsideration of its role in the ecosystem, including a revaluation of the royalty structure with artists, which some perceive as too low; managing rights across global platforms and reissues; and strategies for promoting Japanese music internationally. The success of independents like Kikuo and the global rediscovery of city pop points to untapped potential. The Japanese music industry's future depends on its ability to reconceptualize music in terms of rentier models rather than as a product, and to rethink long-held assumptions about how music should be marketed.

NOTES

1. Ono Tetsutarō, interview with the author, Tokyo, June 13, 2016.

2. Recording Industry in Japan (RIAJ), *Statistics Trends*: 日本のレコード産業 (Japan's record industry), *2024* (Tokyo: RIAJ, 2024), www.riaj.or.jp/f/pdf/issue/industry/RIAJ2024.pdf, 10; International Federation of the Phonographic Industry (IFPI), *Engaging with Music* (London: IFPI, 2023), archived February 13, 2025, at https://web.archive.org/web/20250213055602/www.ifpi.org/wp-content/uploads/2023/12/IFPI-Engaging-With-Music-2023_full-report.pdf, 7; RIAJ, "2023 年度音楽メディアユーザー実態調査 定点調査" (2023 fiscal year music media user survey), March 2024, www.riaj.or.jp/f/pdf/report/mediauser/softuser2023.pdf, 9.

3. See Eric Drott, *Streaming Music, Streaming Capital* (Durham, NC: Duke University Press, 2024), for a theoretical treatment of this argument.

4. Drott, *Streaming Music*.

5. RIAJ, "2018 年度音楽メディアユーザー実態調査 定点調査" (2018 fiscal year music media user survey), March 2019, www.riaj.or.jp/f/pdf/report/mediauser/softuser2018.pdf.

6. Patrick W. Galbraith and Jason G. Karlin, *AKB48* (New York: Bloomsbury, 2019).

7. Only 2 percent of Japanese used an unlicensed music site in 2023, as opposed to 29 percent of global listeners. RIAJ, "2023 年度音楽メディアユーザー実態調査 定点調査" (2023 fiscal year music media user survey); IFPI, *Engaging with Music*.

8. 山口哲一 (Yamaguchi Tetsuichi), 図解入門業界研究 最新音楽業界の動向とカラクリがよくわかる本 (Illustrated introduction to industry research: A book that helps you understand the latest music industry trends and mechanisms) (Tokyo: 秀和システム [Shuwa System Co.], 2022), 46.

9. See RIAJ, *Statistics Trends: The Recording Industry in Japan, 2017* (Tokyo: RIAJ, 2017), www.riaj.or.jp/f/pdf/issue/industry/RIAJ2017E.pdf, 5, for 2016 figures. In 2023, this number was 88.7 percent; see RIAJ, *Statistics Trends: The Recording Industry in Japan, 2024* (Tokyo: RIAJ, 2024), www.riaj.or.jp/f/pdf/issue/industry/RIAJ2024E.pdf, 5.

10. "夏のご褒美カタログ200　定額サービス　定額制音楽配信サービス　『無料で聞き放題』で圧倒的な差別化を図るスポティファイ　日本人アーティストや人気の邦楽ラインナップは限定的" (Summer reward catalog 200: Flat-rate services, flat-rate music streaming services—Spotify aims for overwhelming differentiation with 'listen all you want for free,' but Japanese artists and popular J-pop lineups are limited), *Nikkei Trendy*, August 2018, 70–71.

11. Ono interview.

12. Ono interview.

13. Spotify Technologies S.A., "Form 20-F 2023," February 2, 2024, https://s29.q4cdn.com/175625835 /files/doc_financials/2023/ar/26aaaf29-7cd9-4a5d-ab1f-b06277f5f2a5.pdf, 43.

14. In 2023, Spotify's ad-supported revenues totaled €1,681 million, with a cost of €1,619 million, mainly from royalties. Freemium revenues were 12.7 percent of total revenues. If one allocated this percentage to the €3,843 million Spotify spent on R & D, sales and marketing, and general and administrative expenses, the operating loss on the freemium business was €488 million. Spotify reported a total operating loss of €446 million in 2023. Spotify Technologies S.A., "Form 20-F 2023."

15. A 2020 survey by ICT Research and Consulting (a think tank subsidiary of NTT, the nation's largest telephone company) showed Amazon Prime Music to have the largest user base, at 13 percent of respondents, followed by Spotify (9 percent), Apple Music (9 percent), Line Music (6 percent), Amazon Music Unlimited (6 percent), YouTube Music (4 percent), and AWA (2 percent). A 2023 user survey by digital marketing company Nyle found that Spotify had the largest number of users, with a 23 percent share of accounts, versus Apple at 18 percent, Amazon Prime at 17 percent, YouTube Music at 13 percent, Line Music at 9 percent, and Amazon Music Unlimited at 6 percent; however, these figures would not be indicative of revenues because of differences in the sales mix between freemiums and subscriptions. See Nikkei Value Search, 音楽配信 (Music streaming) (Tokyo: Nikkei, 2022); Nyle, "約半数が音楽配信サービスを利用　10代20代は6割超え　人気は「Spotify」「Apple Music」（Appliv TOPICS調べ）" (Approximately half of the population uses music distribution services, with over 60 percent of people in their teens and twenties using them. The most popular are "Spotify" and "Apple Music" [according to Appliv TOPICS]), *PR Times*, September 6, 2023, https://prtimes.jp/main/html /rd/p/000000398.000055900.html.

16. Nyle, "約半数が音楽配信サービスを利用" (Approximately half of the population uses music distribution services); Tsion Tadesse, "Global Music Subscription Market Sees Growth, Led by Emerging Markets," MIDiA Research, February 8, 2024, www.midiaresearch.com/blog/global-music -subscription-market-sees-growth-led-by-emerging-markets-midia-research-q3-2023-report.

17. Nyle, "約半数が音楽配信サービスを利用" (Approximately half of the population uses music distribution services); IFPI, *Engaging with Music*.

18. United Nations Population Fund, "World Population Dashboard," accessed July 6, 2024, www .unfpa.org/data/world-population-dashboard. Figures for 2024. For similar but slightly different numbers, see Central Intelligence Agency, "The World Factbook," accessed July 6, 2024, www.cia.gov/the -world-factbook/countries/.

19. IFPI, *Engaging with Music*; RIAJ, "2023 年度音楽メディアユーザー実態調査 定点調査."

20. Drott, *Streaming Music*.

21. Tokuda Masashi, interview with the author, Tokyo, January 6, 2010.

22. IFPI, *Engaging with Music*; RIAJ, "2023 年度音楽メディアユーザー実態調査 定点調査" (2023 fiscal year music media user survey).

23. RIAJ, "2023 年度音楽メディアユーザー実態調査 定点調査" (2023 fiscal year music media user survey).

24. 鈴木おさむ (Suzuki Osamu), "１９７０年代生まれの団ジュニたちへ　８３回　無料 で音楽を聴くと罪悪感を感じる世代の楽しみ方　※定額のサブスク（リプションサービ ス）で音楽を聴くようになったおじさんは思う" (To those born in the 1970s, part 83: How to enjoy music for a generation that feels guilty about listening to music for free; thoughts from an older man who started listening to music through fixed-price subscription services), *Shūkan Asahi*, January 25, 2019, 121. Translated from the Japanese by the author.

25. Patrick St. Michel, "Make Believe Mailer 38: Culture Convenience Club," *Make Believe Mailer* (blog), March 25, 2022, https://mbmelodies.substack.com/p/make-believe-mailer-38-culture -convenience.

26. RIAJ, "2023 年度音楽メディアユーザー実態調査 定点調査" (2023 fiscal year music media user survey).

27. Ono interview.

28. Glenn Peoples, "U.S. Music Consumption Saw Double-Digit Growth in 2023 as Streaming Surged, Sales Rebounded," *Billboard*, January 10, 2024, www.billboard.com/pro/2023-u-s-year-end-luminate-music-report-streaming-rises-sales-rebound/.

29. Ono interview.

30. IFPI, *Engaging with Music.*

31. Peoples, "U.S. Music Consumption Saw Double-Digit Growth in 2023."

32. Avex, "Medium-Term Management Plan (FY2023–FY2027)," May 12, 2022, https://contents.xj-storage.jp/xcontents/AS00612/52270394/9b36/490e/b6ca/9e13a837d296/20220527153759850s.pdf; Avex, "Explanation of FY24 Earnings and Avex 2027 Vision," May 15, 2024, https://contents.xj-storage.jp/xcontents/AS00612/4f32f528/22bc/46ec/85fe/b194cb0d1864/20240515161928783s.pdf.

33. Warner Music Group, "Q2 2024 Earnings Conference Call," May 9, 2024, https://edge.media-server.com/mmc/p/8seszuei/.

34. 鈴木スージー (Suzuki Susie), "時代の試練に耐える音楽を――「落ちこぼれ」から歩んできた山下達郎の半世紀（Yahoo!ニュース オリジナル 特集）" (Music that stands the test of time: Yamashita Tatsurō's half century of progress from being a "dropout" [Yahoo! News original special]), *Yahoo!ニュース*, June 11, 2022, https://news.yahoo.co.jp/articles/be612cd888a261a17c38007d9f51406f35ddaade.

35. 山口(Yamaguchi), *図解入門業界研究* (Illustrated introduction to industry research), 16–19.

36. 安藤和宏 (Andō Kazuhiro), "ストリーム配信におけるValue Gap問題に関する一考" (A consideration on the value gap problem in streaming services), *慶應法学* 52 (March 2024): 7.

37. Digital, Culture, Media, and Sport Committee, *Economics of Music Streaming* (London: House of Commons, 2021), https://committees.parliament.uk/publications/6739/documents/72525/default/, 17.

38. Murray Stassen, "The Major Record Companies (and Merlin) Saw Their Spotify Market Share Fall Again in 2023. Yet Even before 'Artist-Centric' Changes, This Decline Started Slowing," *Music Business Worldwide*, February 12, 2024, www.musicbusinessworldwide.com/major-record-companies-and-merlin-spotify/. Figures based on Spotify's annual reports.

39. "Music Stats 2023," TuneCore Japan, accessed June 23, 2024, www.tunecore.co.jp/music-stats/2023.

40. Sikano Mizuki, "きくおインタビュー。「美とは整合性」という哲学とボカロ・ヒット曲「愛して愛して愛して」など楽曲制作の裏側を聞く @ニコニコ超会議2024" (Kikuo interview: We ask about his philosophy of "beauty is consistency" and the behind-the-scenes story of song production, including his Vocaloid hit "Love Me Love Me Love Me" @ Niconico Chōkaigi 2024), *plug+*, May 22, 2024, https://plugplus.rittor-music.co.jp/learning/interview/cho2024_kikuo/.

41. 山口(Yamaguchi), *図解入門業界研究* (Illustrated introduction to industry research).

42. RIAJ, *The Record* 719 (October 10, 2019), www.riaj.or.jp/riaj/open/open-record!file?fid=1785, 3.

43. Tomonori Shiba, "ボカロP・きくおが語る、海外でファンダムを築く方法" (Vocaloid producer Kikuo talks about how to build a fan base overseas), *Rolling Stone Japan*, June 12, 2023, https://rollingstonejapan.com/articles/detail/39548/1/1/1.

8

Inside Playlist Pitching

Music Promotion on Streaming Platforms in Italy

Francesco D'Amato

The promotion of recorded music includes activities that aim to increase awareness and interest among listeners. The achievement of these goals depends in large part on the ways music is presented, categorized, positioned, and contextualized within the points of contact with the public, such as—for example—radio stations, music press, retail stores, and, more recently, music streaming platforms. These institutions act in different ways as gatekeepers and curators, selecting, organizing, and presenting the music to be offered to their audiences, with goals that are often divergent from those of creators and producers. In this way, they not only contribute to the formation of musical audiences and markets but also influence the artistic choices and marketing strategies of producers, who are compelled to consider the criteria and modus operandi of these intermediaries. On the other hand, these intermediaries are, in turn, dependent on people and organizations who produce and provide musical content. Although this interdependence always implies a certain degree of cooperation between the parties, the way this is carried out and the degree and forms of mutual adaptation are expressions of dynamic power relationships. Thus, music promotion has often been shaped by negotiations and tensions between the parties, as well as by tactics, on the part of musicians and labels, to exploit the opportunities and circumvent the constraints posed by the intermediaries.[1]

Today, the chances of getting recorded music known and monetized depend in large part on the selection and curation activities of streaming platforms, which represent not only the main point of access to digital music but also an important locus of music discovery and a potential guide in building listening pathways.[2] Those promoting music on music streaming platforms (MSPs) must therefore take into account the search and recommendation mechanisms implemented by the platforms. Since the mid-2010s, the focus of MSPs' curatorial activities has consisted of creating and promoting playlists, a device of great significance in directing users' listening. Some of these playlists are generated through algorithms, while others curated by the editorial staff, who may, however, employ automated search and recommendation mechanisms. The inclusion of tracks in MSPs' playlists, particularly in editorial ones, as well as the type of inclusion—that is, in which and how many playlists and in what position—can contribute significantly to the exposure and monetization of those tracks. Playlist curation thus represents the main expression of the power of mediation exercised by MSPs toward the music industry.[3] Music streaming promotion can thus include various "optimization" techniques aimed at influencing the discoverability of a musician or a song via user searches and algorithmic recommendations, playlist pitching, and payment for services that offer inclusion in popular playlists and an increase in streams.[4]

To the extent that studies of the platformization of cultural production are concerned with the ways in which the economic logics, technological infrastructures, datafication, and curation activities of digital platforms influence the processes of creation, marketing, distribution, and monetization,[5] analyses of the curatorial activities of MSPs, and of the tactics developed by music creators and producers in response to them, can be understood as contributions to the study of platformization in the music field. So, too, can the implications of these processes for the visibility and monetization of different music projects and musicians. At the same time, studies of platformization should also recognize that relations between producers and platforms cannot be reduced to dependence of the former on the latter, but rather must be framed in terms of an interdependence that is configured through mutual adaptations—sometimes more collaborative, sometimes more conflictual—within asymmetrical power relations.

This chapter offers a complementary contribution to studies of music streaming and platformization in the field of music by examining *playlist pitching*. In particular, it analyzes the constraints experienced and the criteria of action adopted by those who seek to include music in successful playlists, as well as their relationships and negotiations with their counterparts in MSPs. Such analysis therefore differs from studies focused more on algorithmic playlists,[6] playlists' relevance for MSPs' business strategies and the contents they favor,[7] the purchase of playlist placements through paid services,[8] or the work of MPSs' curators.[9] Its aim is to detect the factors that influence playlist pitching and its effectiveness, as well as their implications for different types of musicians, labels, and music projects. In addition, elements of continuity and discontinuity from more traditional forms of media promotion will be highlighted.

The analysis presented is the result of research conducted in the Italian context, through twenty-four in-depth interviews, each lasting at least one hour, with sixteen professionals (some of whom were interviewed twice to compare and deepen information that had emerged from the other interviews). Each participant was involved in many music streaming promotion activities, including playlist pitching, comarketing initiatives with MSPs, curation and promotion of playlists for their organization, and the use of social media and influencers. The interviewees held different roles within different types of organizations: two heads of digital (HoD) and five digital account managers (DA) from the Italian divisions of two different majors; one general manager (GM) and three marketing managers (MM) from the Italian divisions of two different international digital distributors; one general manager of an Italian digital distributor; three digital promotion managers (DPM) within important Italian independent labels; and one international development manager for a major label.[10]

MUSIC STREAMING IN THE ITALIAN CONTEXT

In Italy, music streaming started to grow significantly around 2013, the year in which Spotify entered the market, joining Deezer (2012), Cubomusica (launched in 2011 by Telecom Italia), and other minor MSPs. In 2014, streaming surpassed downloading, and by 2023, music streaming had come to represent 65 percent of the recorded music market (the third largest within the European Union).[11] By this time, on Spotify alone, more than 1,200 Italian musicians generated over €10,000 in royalties, a number that has more than tripled since 2017. Furthermore, 50 percent of all royalties generated by Italian musicians (€126 million, up 400 percent compared to 2017) came from users based outside Italy.[12]

The growth of streaming has coincided with the expansion of the market for Italian music. Will Page and Chris Dalla Riva found that in 2022, Italy was the only European country in which the top ten artists were all "local" and the nation in which the percentage of top ten tracks by local musicians was highest (70 percent, on par with the United Kingdom, Sweden, and Poland).[13] By 2023, 80 percent of the songs in Spotify's "Top 50 Italy" playlist were by Italian musicians,[14] as were 80 percent of the top one hundred albums in general (a figure that stood at 63 percent in 2013).[15] However, contrary to what Page and Riva seem to maintain, it is problematic to attribute these results solely to streaming, since the prevalence of local repertoire in Italy preceded streaming by about ten years.[16] The interpretation given by some interviewees—that streaming has not revolutionized the local market but has adapted to and amplified already existing trends—seems therefore more correct.

Playlists seem to play a crucial role in this context: according to Spotify, more than half of the discovery of new artists occurs through editorial or algorithmic playlists, and in 2023, more than five thousand Italian musicians were included in its editorial playlists.[17] It is, therefore, understandable that playlists are considered crucial for early career development, despite several interviewees downplaying their importance. However, as we shall see, this argument is mostly used by labels and distributors with musicians who complain about not being included in playlists.

PRESELECTION AND HIERARCHIZATION

Music promotion activities that involve MSPs, in particular playlist pitching and comarketing initiatives, are mainly carried out by the digital departments of the majors and by digital distributors, in collaboration with the labels. Few MSPs allow even self-distributing independent musicians to pitch their own music; however, this chapter will focus on the activities of major labels and digital distributors. Which promotional activities can be carried out, and in what manner, depends first and foremost on whether an MSP has local staff, as well as on its organization and resources. The absence of editorial staff implies a lack of locally curated playlists, which generally means less attention to both local repertoires and specific local market preferences and trends. The only MSPs with editorial staff in Italy as of April 2024—and therefore considered the most relevant by interviewees—were, in order of importance, Spotify, Amazon Music, and Apple Music. There is no public data on Italian streaming market share among the different MSPs, however respondents agree that at least 70 percent is held by Spotify. The local divisions of these three MSPs include people in the role of *label relations*, who deal most with labels and distributors, particularly for comarketing, social media, and in-platform promotion, and people in the role of *music curation* (or *programming*), who create and develop editorial playlists. In Italy, Spotify has two curators and two label relations, Apple Music has one person for each of the two areas, while Amazon Music has a larger and more diverse structure, including three curators. As of April 2024, there were ninety-four playlists curated by Italian editors on Spotify, fifty-seven on Amazon Music, and thirty-nine on Apple Music.

The main activities of playlist pitching consist of compiling and sending to the MSPs the *release schedule* (RS), *pitches* using automated tools (for the MSPs that have them—again, Spotify, Amazon, and Apple), and *repitches*. To these must be added *direct calls*, mostly to label relations staff, and *listening sessions* dedicated to the most important projects, in which curators also participate.

The RS consists of the list of upcoming local and international releases flagged as priorities for locally curated playlists, differentiated in turn by priority level, as in traditional media promotion. Most interviewees agreed that the level of priority significantly affects the chances of achieving (satisfactory) playlisting. These chances vary by MSP, in relation to aspects discussed in the next section. The releases included in the RS, generally between twenty and forty, tend to be mostly Italian, relying on the distinctiveness of the local market, characterized by a predominant share of domestic repertoire. The receptivity of MSPs to local proposals is thus dictated by the context, as pointed out by the HoD of major A:

"If they want to grow in Italy, they have to favor local repertoires," in both the composition of playlists and the choice of projects on which to activate comarketing operations.

The selection and prioritization of new releases for inclusion in the RS are at the discretion of local teams of majors and distributors, except for international priorities imposed by central offices. These operations are carried out with consideration of the following: labels' indications regarding their internal priorities, metrics related to musicians' performance, the playlist ecosystem of individual MSPs, and major label and distributor staff members' beliefs about the preferences and reasoning that guide MSP curators' choices. The number of weekly priorities is determined according to two criteria: to avoid as much competition as possible among releases from the same major/distributor and to maximize playlisting chances for the greatest number of them, considering the opportunities identified on the different platforms. "As these opportunities increase, that is, the number of playlists that constitute possible landings, the number of priorities we enter into RS also increases" (HoD, major B). A digital account manager of a different major notes how it is "pointless to do the work on fifty things if you already know that they will accommodate twenty." Similarly, the choice of songs to be included, and especially their order of priority, takes into account the differences between MSPs in terms of locally curated playlists and curators' modus operandi. Songs that fall into musical genres for which there is a lack of locally curated playlists (e.g., black metal, phonk, fusion), when included in the RS, are accompanied by a request for Italian curators to report them to their colleagues who curate international playlists for those genres. This implies that, on the one hand, it is easier for songs by Italian musicians who play these types of music to be *proposed* for inclusion in international playlists—unlike, for example, those of Italian pop or rock musicians—since, in the absence of local playlists, their pitching is inevitably directed to international ones; on the other hand, *actual entry* into these playlists is more difficult due to greater competition. For these reasons, tracks ascribed to genres that are not curated, or that are poorly curated, by the local MSPs staff tend to be penalized in assessing the priorities:

> If we are notified [by the labels] of an indie folk song sung in French by an Italian [performer], we already know that the spaces that the *partners* [i.e., the MSPs] objectively devote to that kind of music are extremely narrow; therefore, it is better not to give it the highest priority so as not to sacrifice that priority slot at the expense of someone else. . . . We try to prioritize the songs that are most placeable in the platforms. (DA1, major B)

> If we know that a specific genre is not curated by local curators, we have a harder time pitching it. (DA3, major A)

At the same time, the outcome of these selection and hierarchization processes depends, in part, on contingent situations, such as the amount and type of weekly releases:

> For example, this week we have a lot of releases, but we don't have any super big musicians, locally or internationally. So, in the second or third place, I have a chance to place an emerging artist. So, this week is favorable; if it had been any other week, the same artist who is in third place could have found himself in fifteenth place. (DA2, major A)

FINDING A PLACE IN THE ECOSYSTEM (OR FORCING ITS EXPANSION)

While sending the RS is the only way to report priorities to MSPs that do not have a specific pitching tool, in other cases the latter is also used. Unlike the RS, which includes both local and international priorities, pitching via the tool is possible only for Italian productions. For songs in the RS, the pitch is usually made by DPMs or DAs handling digital distribution, based on information provided by the labels' A&Rs and artists' managers, while the pitch for tracks that do not enter the RS is often made directly by the labels. The goal of the pitch is to ignite the interest of the curators and guide their framing of the song, so that it will be placed in the most coherent and relevant playlists. The tools allow the insertion of various types of information, such as genre, mood, style, instruments, language, and so forth, mostly constrained by the selection of options within predetermined menus. The type of information required by the different tools is similar, but the options and choices are not (e.g., Spotify allows three genres to be indicated from more detailed lists than the others). Amazon's tool (Maestro) allows users to indicate a playlist that is considered "ideal" for the song, whereas the other two do not; however, while Apple still allows users to indicate the desired playlist in the RS, "at Spotify, they are more restrictive on this; they value their freedom a lot, though they would be free anyway because I could suggest a playlist, but then the choice is still up to them; I'm not imposing anything; I just wish there was more exchange on these things, but they don't want to" (MM, distributor B). The fact that MSPs, especially Spotify, often fail to notify DAs and MMs about the creation of new playlists further highlights their desire to carry out the curatorial work without worrying about its relevance to music promotion activities.

The pitching tools also allow for the addition of a short text, widely considered critically important for stimulating the interest of curators and encouraging greater attention to the song—a fact many believe is not to be taken for granted, given the overall volume of weekly releases. The information considered most effective for this purpose falls into two types: on the one hand, those capable of enhancing the project and the musician on a cultural level, such as the status of the authors and producers, or the musicians featured in the song or those responsible for discovering the promoted artist; on the other hand, those concerning plans for medium-term development (i.e., the schedule of subsequent releases and tours). According to interviewees, showing MSPs that the release is part of a broader project serves to convince them that fostering its visibility and growth would benefit them as well, in terms of listeners interested in streaming subsequent releases. This is the same kind of argument traditionally used with radio stations and retailers as well. A third category of information, particularly relevant for the first releases of new artists, consists of the metrics that measure performance on social media. While these factors coincide in part with those influencing the initial playlisting choices identified by Arnt Maasø and Anja Nylund Hagen, Benjamin Morgan also detected the potential weight of curators' reliance on promoters who had previously repeatedly proposed tracks that performed well after being playlisted.[18]

In addition to being included on RSs and pitched via automated tools, the most important releases are presented in calls with label relations staff at MSPs, who receive more detailed information on planned marketing activities and, in turn, will talk to the curators about those releases. Furthermore, majors and distributors hold periodic listening sessions on main projects, also involving A&Rs and musicians, well in advance of the releases and with curators in attendance.

At the end of these processes, there may still be instances of noninclusion in playlists, inclusion in playlists deemed to be inadequate (poorly followed), or inclusion in playlists misaligned with respect to the musician or project, especially for low priority tracks. The causes are generally traced to various combinations of three factors: the curators' evaluations, their listening dispositions and criteria of classification, and the limitations of locally curated playlists in the ecosystem (*ecosystem* is the term always used by interviewees). These limitations concern not only the *quantity* of playlists curated locally by different MSPs but also their *rigidity*. Some of the DAs interviewed have pointed out that although Apple has fewer playlists, some of them are very loosely defined, and so can accommodate heterogeneous tracks; in contrast, Spotify has many playlists with greater internal consistency, which limits the possibilities for inclusion of stylistically divergent tracks. Any mismatch between the genre in which labels frame the music project, or in which the artist recognizes themselves, and the genre of the playlist in which it is placed, is considered the least serious problem. One example involves musician M, who identifies as a singer-songwriter and is promoted as such, but he found himself placed in the indie pop playlist—not because there is no playlist dedicated to songwriting, but because that playlist primarily focuses on "classic" Italian songwriters of the past and stylistically similar contemporary productions. Curators felt that M's music aligned more with the indie pop style than that of the "classic" songwriters. Generally, no countermeasures are taken in these cases, either because most interviewees say they more often agree with MSPs' classification choices than with label indications, or because, according to respondents, labels and musicians tend to care more about the economic advantages and symbolic recognition brought by placement on a good playlist than about genre fit. However, in these cases, what is important for the DAs is to understand "how the curator sees the musician, so [that] next time we position him/her that way" (DA2, major A).

The lack of playlisting may result from a negative evaluation of the song by the local curators or—if deemed unsuitable for the locally curated playlists—by the foreign curators of the playlists in which it could have found appropriate placement. The ecosystem of local playlists can, therefore, be inhospitable to certain types of productions, forcing them to seek space in places where the competition is greater. An example given by two different DAs concerns the type of music that both defined as "alternative pop," which is too far removed from the songs that make up Italian indie pop and alternative rock playlists. The limitations of local curation can also penalize foreign songs that are excluded from international playlists but could be profitably promoted in the Italian market if there were a version of those playlists calibrated to the tastes of Italian audiences: "I've been asking for eight years to make an 'international rock' playlist run from Italy because the alternative music we listen to in Italy is different from the alternative music they listen to abroad. Even some types of dance music perform differently depending on the country" (DA1, major A). This comment exemplifies a view shared by all interviewees: the issue of the absence of playlists that can accommodate particular repertoires, which are of strong interest to labels and distributors, cannot be resolved by simply suggesting their creation to MSPs. New playlists are created only when the MSP *autonomously* detects a new consumer trend it deems worthy of a playlist. One tactic employed by the majors to indirectly solicit the creation of editorial playlists dedicated to their target genres is to leverage their own playlists to promote those repertoires. For example, speaking of the emerging Italian Afropop music scene, one major's digital account manager explained that

> in that case we created a playlist ourselves called [name omitted], and we launched it with an event and the release of a press note, getting it out to MSPs as well, as if to say, "We in the meantime move and create a pool of listeners that we try to grow with our resources; however, we point out to you that, for us, it is an absolutely relevant and growing pool." With our brand of playlists, we have extra valorization opportunities, we have budget to invest for in-platform advertising, we have social channels that we can exploit in an organized way. We have a firepower that allows us to become a reference for the Afropop scene within the platform. (DA2, major B)

> We discover genres that no one considers, and maybe we start making a playlist of that genre before someone else does, so we get indexed ourselves, then maybe we sign artists of that genre because we see that it is working, and we also already have the number one playlist of that genre. (DA1, major A)

Labels then use their own playlists to test or push emerging genres that have not yet attracted the interest of curators.

"ACTIVATING" THE AUDIENCE

In cases where the failed or unsatisfactory playlist placement depends primarily on value assessments, a repitch can be tried. This takes the form of an email directed to the curators, intended to bring to their attention data and arguments supporting a request: to playlist a song not included in any playlist, move it to more important playlists, or include it in additional playlists. Although, in principle, anyone with access to curators' email addresses could submit repitches (though these are not easy to obtain), majors and distributors point out that good repitching needs more than just contact information. It also demands knowledge regarding what information MSPs consider most relevant, access to additional data beyond what is readily available (e.g., through Spotify for Artists), and the ability to integrate and process this data, which requires expensive technology. As with RS, and unlike pitches made with pitching tools, repitches can involve Italian or international songs proposed for locally curated playlists. The metric considered by far the most relevant—and on which, according to some, a song's very "repitchability" depends—is not the growth of overall streams but rather the increase in the percentage of streams defined as "active." These are streams coming from organic searches on MSPs, visits to the artist's profile, plays from a listener's library, or clicks on external links, such as those included in posts published on social media to promote a new release. In other words, they encompass all listening that does *not* come from playlists curated by the MSP or others. Listening to a playlist is thus considered a passive reception of music proposed by others. According to several interviewees, there are even repitchability thresholds based on this metric—though not formalized, they are the result of convention and are especially relevant for Spotify and Amazon, as these quotations show:

> A song can only be repitched when active listens are much greater than passive listens, as it means that the song is not performing well just because it is in the playlists. On the other hand, when passive listening rates are higher than a certain threshold, it means that the song already has enough playlists, and we cannot ask for more. (DA2, major A)

> The minimum organic traffic that the song must have for an effective repitch, or at least for it to be repitchable, goes from a minimum of 40 percent and up; if you have less than 40 percent organic you don't even repitch it, because you already know they're never going to put it on a playlist anyway. (DA1, major B)

In the words of one distributor's marketing manager, by demonstrating the degree of actual interest in a song, organic listens attest to its value for editorial playlists. Repitchability thresholds may be negotiated depending on the project. For example, the repitch of a highly successful pop musician's song, which "by default" ends up in highly listened-to playlists and thus "inevitably" has many passive streams, can be positively received even with relatively low percentages of active streams. One consequence of the lesser weight given to streams from playlists is that a significant part of playlist-promotion activities focuses on engaging audiences directly, encouraging them to listen to the song on the platform. Initiatives used for this purpose fall within the areas of social media marketing, guerrilla marketing, and influencer marketing. However, interviewees' insistence on the involvement of musicians interacting through their social media is striking:

> We teach artists that they don't have to worry about talking to Spotify curators; they have to talk to fans. They have to get them to listen to them on that platform—through stories, posts, putting the link to the platform, making up whatever they want, marketing activities that we put money on. . . . You have to show [MSPs] things with metrics. If the artist is committed to getting his fans to listen to his music there, all the metrics increase, so then we can say to MSPs, "Look, this is getting better; you should reconsider this." (DA1, major A)

> I always say, "Let's make things happen, then Spotify notices." That's the key to getting into this virtuous circle—making something happen for your fan base. It is fundamental how the artist tells himself on social media, with TikTok, Twitch. Today's emerging artist must have these prerogatives; otherwise, he's making it three times harder—or not making it at all. (DP, indie A)

In such cases, the use of labels' playlists to push the nonplaylisted track risks being counterproductive. While it may provide an opportunity to have the track discovered and "saved" in users' libraries, possibly fueling "active" streams, it also risks increasing playlist listens and, thus, generating "harmful" metrics for repitch purposes. In addition to "active" streams, the other two notable streaming metrics are user saves, also considered evidence of concrete interest in the song, and, to a lesser extent, skip rate, considered an "active" manifestation of disinterest or dislike. Repitch recipients are also sensitive to data regarding the performance of the song on the other MSPs.

It is not just streaming data that is considered useful for repitch objectives. Many interviewees insisted on the importance of data related to both searches for the song on the music-identification service Shazam and the eventual growth of its uses in user-generated content on TikTok. Shazam searches seem to count more than data pertaining to radio airplay, precisely because, even in this case, they are considered a manifestation of actual interest, rather than potentially distracted listening. Regarding viral phenomena on TikTok, which often involve songs released several years earlier, this data can be useful for repitching catalog songs, even long after their release. However, interviewees' choice and use of data are not determined entirely by the logic and functioning of the MSPs; they also derive from the tactical ability of promoters to detect and propose interpretations that serve their goals. One of the directors of the digital department of a major was keen to point out that

> all this stuff of active versus passive is more of an internal convention, which we need to identify an "objective" argument to use toward the *partners* [the MSPs], [which is] different from saying, "In our opinion this track should be there," which we can't do with them. . . . It's a working convention, but then on individual tracks, you make qualitative distinctions. So, if you have a thousand streams, you base the repitch on completely different arguments. For example, you base it on the fact that it's a musician we're investing in; it's a priority, and so you develop more of a partnership discourse, which is something you can't do with the curator, but you do it with the label relations people, who then convey the message to the editorial team anyway. . . . If, instead, I have an Afrobeat track that generates 90 percent of streams from a taste-maker's playlist, the repitch will be based on this—not distinguishing between active and passive streams, but highlighting the fact that it is placed within a tastemaker's playlist. (HD, major B)

ELUSIVE PARTNERS

The design and effectiveness of music promotion depends in large part on the knowledge and understanding of the interlocutors. While choices regarding playlist pitching can rely on a rough knowledge of the general processes and principles structuring curatorial activities, explanations about the actual playlisting decisions regarding a particular song, in the absence of timely feedback from curators, are based on inferences. These, in turn, rest on observing decisions made each week by the curators on a large number of releases, on continuous indirect contacts (the interactions with label relations, who, according to the interviewees, confront the curators about their respective decisions), and on the more rarefied direct contacts at listening sessions and other informal meetings. Limitations in interactions with partners (a term consistently used by interviewees in reference to MSPs) and, therefore, constraints in understanding their curatorial choices, as well as the centrality of data, emerge as peculiarities of streaming promotion—particularly through comparisons with "traditional" promotion in other media, especially radio.

Beyond the respondents' considerations, a first difference between radio and streaming promotion can be attributed to the degree of concentration of the music streaming industry. An important aspect of traditional media promotion is "matchmaking": the identification of channels and people deemed most in line with the projects to be promoted and, thus, potentially most receptive and best disposed toward them.[19] In the current Italian context, where playlist promotion is aimed at only three MSPs and six curators, with whom there are almost no opportunities for direct confrontation, matchmaking hardly exists. The small number of interlocutors, together with the huge number of new releases in recent years,[20] would still make it complicated, according to the interviewees, to obtain accurate feedback on their choices—even if they had a different policy and there was more willingness to engage in dialogue, such as that which characterizes the relationship with radio stations instead.

> [With radio,] there is still a very personal relationship. The radio person has to be cuddled a little bit; the curator doesn't have to be cuddled because, quite simply, he doesn't even respond to you. . . . While with the label relation there is really an exchange of ideas, the curators read all the emails and pitches, but then they don't give us feedback; they don't tell you, "Well, this project isn't suitable for that particular playlist for this reason." Before [i.e., until the mid-2010s] there was a little more exchange; now communication is one-way. (DP, indie A)

> The curators are kind of untouchable, in the sense that we don't interface much with them. We report weekly priorities, but then we don't have a dialogue about what they playlist or not; communication with them is one-sided. (DA2, major A)

> [Radio] is something that is easier to have a dialogue with; with MSPs, there is always a bit of a wall. . . . Also, you never really know what the strategy or purpose of an MSP is for what concerns music. (GM, distributor A)

The theme that frequently emerges in the interviews concerns the stubbornness of MSPs (some more than others) to evade confrontation over curatorial activity in the name of preserving their autonomy. In fact, this attitude can also be found in some radio stations, as evidenced by research conducted in other contexts.[21] The lack of feedback on playlisting choices poses a problem, in part analogous to that of trying to understand the functioning of algorithms: the formation of knowledge to enable, on the one hand, effective adaptation of promotional tactics to the context to which they relate, and, on the other hand, accountability for (un)achieved results to the other stakeholders involved—in this case, labels and musicians. At the same time, the discretion of majors and distributors—in the selection and hierarchization of priorities, the final formulation of pitches, and in choosing the arguments for repitches—is legitimized by reference to their superior knowledge of MSPs, compared, for example, to that of smaller labels. Such inferred knowledge results from occupying a privileged observation point, from the greater frequency of direct and indirect contacts, as well as from the rare feedback received in exceptional cases.

> It is mostly Spotify curators who do not expose themselves on their ratings and insertions, while Amazon and Apple are a bit more open to confrontation, at least to make me understand a minimum, so that I can report to colleagues or musicians their reasoning [and] explain why the song was not put there. . . . The easiest of all is Apple, with whom there is a very serene human relationship; we know each other very well, also, because we have been working with them for years now. With Spotify, only in truly sensational cases [of lack of playlisting] is the confrontation

opened a bit; it happened that for very important cases, we pointed out the problem to them and they responded. . . . With Amazon, it depends on who is the curator of that playlist; there are curators that maybe you happen to meet at concerts, et cetera, that you get a little bit more familiar with, so maybe he allows you to ask him things and allows himself to expose himself a little bit. Maybe in some situations, you can even say to him, "Look, there might be a problem with the artist, so let me understand a little bit, so we can deal with it and proactively avoid it for you as well." (DA1, major B)

It also happens that we meet in less formal situations; for example, at release parties for a new album, you invite the MSPs, and so there is a way to talk to them, make a little joke, and ask them something with a drink in hand. (DA3, major A)

In this situation of rarefied confrontations and difficulty in presenting projects in a more articulate way, data assume a central role in regulating the intertwining of promotional and curatorial activities. The fact that, for some interviewees, this is a *positive* fact, as also noted by Maasø and Hagen,[22] points to the relevance of accounting knowledge in situations of uncertainty,[23] such as the development and justification of actions addressed to interlocutors who do not provide explanations regarding their choices. In these situations, data are a resource to draw on to satisfy the need for certainty and simplification, providing an "objective" basis for one's choices and internal confrontation, although they merely illustrate performances whose explanation remains largely open to various hypotheses.

We say to the artists, "OK, you are not satisfied; let's wait a week and see the data." That's the great thing about platforms compared to radio, where, instead, there is a continuous negotiation based on nothing, whereas here, you say, "Let's see the performance, and if it's good, let's report and try to get more consideration." . . . The great thing about streaming, for me who studied economics and has always been very analytical, is that you can count everything, so you always have a numerical basis for all the talk. You can develop some work and justify the demand for attention from [the MSPs] toward one artist, based on objective numbers. (DA2, major A)

When the [label staff] tell us, "Why isn't it on this playlist? We're investing in it, we're promoting it on the radio, this and this," it comes in handy to be able to say, "OK, it's true that you're doing everything you can for this song, but if you look at the objective data, you can see that the feedback is very different." (DA2, major B)

CONCLUSION

Although, in the context of streaming promotion, playlist pitching activities are most similar to traditional forms of media promotion, their analysis nonetheless offers insights into the concrete modes of interaction that affect both the perception of MSPs by music providers and the relationships and mutual conditioning among them.

To the extent that MSPs' playlisting choices are considered relevant to the opportunities for visibility and monetization of music content, activities aimed at promoting favorable playlisting must necessarily adapt to MSPs' logics and ways of organizing music. These curatorial logics, and consequently the ways in which promotion is carried out, tend to penalize music productions that differ from locally curated ones, or rather, to generate different types of difficulties for different repertoires. In addition, promotion relies heavily on the information provided by the platforms themselves, which represents only part of the information available to them.

On the other hand, dependence does not determine actions, and adaptation is not passive adjustment. To the extent that MSPs must adapt to market characteristics and trends, the market power of some actors and their ability to act on it can be resources for developing tactics to counteract MSPs' constraints. Of course, this implies that not all tactics can be adopted (especially with the same effectiveness) by any player. A more general implication is that the outcomes of the power relations between producers and platforms depend, in part, on the ways in which audiences react to different solicitations coming from both.

A similar argument can be made in relation to data. Other research has pointed to the centrality of streaming and social media data for the development of artistic and marketing strategies of musicians, managers, labels, and distributors. On the one hand, this is considered an aspect of platformization, evidence of the power of platforms to produce the information on which the activities of other stakeholders depend; on the other hand, this dependence does not determine stakeholder practices. This is not only because it differs in degree and form, depending on the resources and skills that different stakeholders are able to mobilize for the use of the data, but also because these are subject to sensemaking—the critical evaluation of their different possible interpretations and uses.[24] The analysis of playlist pitching helps highlight how this critical work on data, as well as its tactical framing, is also explored in dealing with MSPs, influencing their decisions despite the narrow margins of negotiation, which in turn vary depending on the weight of different stakeholders.

The analysis above helps to clarify how the outcomes of playlist curation derive from the specific ways the interdependence between MSPs and music providers (labels and distributors) is concretely articulated in daily working routines. At the same time, majors and, to some extent, the large international distributors, seem to enjoy greater relationship opportunities and negotiating power with MSPs than other content providers, which may help explain the predominance of their content in the most coveted places and positions.[25]

NOTES

1. Keith Negus, *Producing Pop: Culture and Conflict in the Popular Music Industry* (London: Edward Arnold, 1992), 101–33.

2. Luis Aguiar and Joel Waldfogel, "Platforms, Promotion, and Product Discovery: Evidence from Spotify Playlists," NBER Working Paper w24713 (June 2018), https://ssrn.com/abstract=3198015;

Arnt Maasø and Hendrick Storstein Spilker, "The Streaming Paradox: Untangling the Hybrid Gate-keeping Mechanism of Music Streaming," *Popular Music and Society* 45, no. 3 (2022): 300–316; David Hesmondhalgh, Richard Osborne, Hyojung Sun, and Kenny Barr, "Music Creators' Earnings in the Digital Era," Intellectual Property Office Research Paper, last revised April 27, 2022, https://dx.doi .org/10.2139/ssrn.4089749.

3. Aguiar and Waldfogel "Platforms, Promotion, and Product Discovery"; Tiziano Bonini and Alessandro Gandini, "'First Week Is Editorial, Second Week Is Algorithmic': Platform Gatekeepers and the Platformization of Music Curation," *Social Media and Society*, 5, no. 4 (2019), https://doi .org/10.1177/2056305119880006; Maria Eriksson, "The Editorial Playlist as Container Technology: On Spotify and the Logistical Role of Digital Music Packages," *Journal of Cultural Economics* 13 (2020), 415–27; Robert Prey, "Locating Power in Platformization: Music Streaming Playlists and Curatorial Power," *Social Media and Society* 6, no. 2 (2020), https://doi.org/10.1177/2056305120933291; David Hesmondhalgh, Raquel Campos Valverde, D. Bondy Valdovinos Kaye, and Zhongwei (Mabu) Li, "The Impact of Algorithmically Driven Recommendation Systems on Music Consumption and Production: A Literature Review," UK Centre for Data Ethics and Innovation Reports, 2023, https://ssrn.com /abstract=4365916.

4. Regarding "optimization" techniques, see Jeremy Wade Morris, "Music Platforms and the Optimization of Culture," *Social Media and Society* 6, no. 3 (2020), https://doi.org/10.1177/2056305120940690; and Benjamin A. Morgan, "Optimization of Musical Distribution in Streaming Services: Third-Party Playlist Promotion and Algorithmic Logics of Distribution," in *Rethinking the Music Business: Music Contexts, Rights, Data, and COVID-19*, ed. Guy Morrow, Daniel Nordgård, and Peter Tschmuck (Cham, Switzerland: Springer, 2022), 171–88. Regarding playlist pitching, see Benjamin A. Morgan, "Revenue, Access, and Engagement via the In-House Curated Spotify Playlist in Australia," *Popular Communication* 18, no. 1 (2020): 32–47; and Arnt Maasø and Anja Nylund Hagen, "Metrics and Decision-Making in Music Streaming," *Popular Communication* 18, no. 1 (2020): 18–31. Regarding services for purchasing streams and playlist placement, see Eric Drott, "Fake Streams, Listening Bots, and Click Farms: Counterfeiting Attention in the Streaming Music Economy," *American Music* 38, no. 2 (2020): 153–75; and Morgan, "Optimization of Musical Distribution."

5. Thomas Poell, David Nieborg, and Brooke Erin Duffy, *Platforms and Cultural Production* (Cambridge: Polity Books, 2022), 4.

6. Robert Prey, "Knowing Me, Knowing You: Datafication on Music Streaming Platforms," in *Big Data und Musik: Jahrbuch für Musikwirtschafts- und Musikkulturforschung 1/2018*, ed. Michael Ahlers, Lorenz Grünewald-Schukalla, Anita Jóri, and Holger Schwetter (Cham, Switzerland: Springer, 2019), 9–21.

7. Prey, "Locating Power in Platformization"; Robert Prey, Marc Esteve del Valle, and Leslie Zwerwer, "Platform Pop: Disentangling Spotify's Intermediary Role in the Music Industry," *Information, Communication and Society* 25, no. 1 (2022): 74–92.

8. Morgan, "Optimization of Musical Distribution."

9. Bonini and Gandini, "'First Week Is Editorial.'"

10. The main limitation of this work was that it proved impossible to interview Italian MSP staff, due to explicit refusal or lack of response to the request. To comply with the request for anonymity, the quotations indicate only the role and type of organization, references to specific musicians or musical projects have been removed. All interviews were conducted in Italian, with translations done by the author.

11. Here, I summarize data from several reports released by FIMI (Federation of Italian Music Industry), available at www.fimi.it/mercato-musicale/dati-di-mercato/.

12. Vincenzo Prisco, "Spotify paga 126 milioni all'industria musicale italiana," *Il Sole 24 Ore*, May 16, 2024, www.ilsole24ore.com/art/spotify-paga-126-milioni-all-industria-musicale-italiana-AFoAlR1D.

13. Will Page and Chris Dalla Riva, "'Glocalisation' of Music Streaming within and across Europe," London School of Economics, EIQ Paper 182 (May 2023), www.lse.ac.uk/european-institute/Assets /Documents/LEQS-Discussion-Papers/EIQPaper182.pdf.

14. Prisco, "Spotify paga 126 milioni."

15. FIMI, "Miglior risultato di sempre," 2023, www.fimi.it/mercato-musicale/dati-di-mercato/musica-il-mercato-discografico-italiano-sale-del-18-8-nel-2023.kl.

16. Summarized from FIMI reports, available at www.fimi.it/mercato-musicale/dati-di-mercato/.

17. Prisco, "Spotify paga 126 milioni."

18. Maasø and Hagen, "Metrics and Decision-Making in Music Streaming"; Morgan, "Revenue, Access, and Engagement."

19. Negus, *Producing Pop*, 101–33.

20. Several interviewees underlined how the first months of 2024 marked a significant reversal of this trend, after many years of continuous increase in new releases.

21. Jarl A. Ahlkvist and Robert Faulkner, "Are They Playing Our Songs? Programming Strategies on Commercial Music Radio," in *The Business of Culture: Strategic Perspectives on Entertainment and Media*, ed. Joseph Lampel, Jamal Shamsie, and Theresa K. Lant (New York: Psychology Press, 2005), 155–76.

22. Maasø and Hagen, "Metrics and Decision-Making in Music Streaming."

23. Keith Negus, *Music Genres and Corporate Cultures* (London: Routledge, 1999), 51–52.

24. Maasø and Hagen, "Metrics and Decision-Making"; Nancy Baym et al., "Making Sense of Metrics in the Music Industries," *International Journal of Communication* 15 (2021): 3418–41.

25. Prey, Esteve del Valle, and Zwerwer, "Platform Pop"; Hesmondhalgh, Campos Valverde, Kaye, and Li, "Impact of Algorithmically Driven Recommendation Systems," 18.

9

Changes and Continuities in the Indian Nonfilm Recorded Music Industry under Platformization

Aditya Lal

INTRODUCTION

For decades, the Indian recorded music industry has been dominated by film soundtracks. As Hindi emerged as the putative national language and the Hindi film industry (colloquially known as Bollywood) grew, the musical soundscape of the country became synonymous with film music, especially Bollywood music.[1] The cultural hegemony of Bollywood music was cemented on a national scale by radio and television broadcasting, and the internet further fueled this culture globally, with entrepreneurs in India and abroad launching digital media companies centered on Bollywood films and music.[2]

However, the emergence of domestic and international music streaming platforms (MSPs) in India has significantly changed the status of nonfilm music[3] in that country. For the Indian subsidiary of Spotify, the largest MSP in the world, the most streamed song in 2023 was a nonfilm song.[4] Furthermore, music in regional Indian languages that does not enjoy mainstream exposure through films has also been gaining traction via nonfilm releases.[5] Leading this trend is the Punjabi music sector, where 90 percent of consumption is contributed by nonfilm music.[6] Although contemporary and catalog Bollywood music continue to dominate music consumption in India,[7] Spotify India shows that nonfilm music was growing at a faster rate than film music, now accounting for 10 to 30 percent of Spotify listening.[8]

This makes the Indian nonfilm recorded music industry an interesting site for examining the impact of platformization in a recorded music industry of the Global South. This chapter explores and analyzes the dynamics of the Indian nonfilm recorded music industry in the predigitalization and platformized eras. The next section examines the rise and fall of nonfilm popular music before digitalization, especially focusing on the 1990s, when nonfilm music peaked alongside the advent of transnational satellite television in India. The section after that explores the current state of nonfilm music under platformization and highlights the distinctiveness of the Indian music streaming market. This is followed by an analysis of the changes effected (or not) by platformization in the Indian recorded music industry, along with a commentary on how developments in media technologies transformed the Indian recorded music industry while retaining some important key features. The chapter argues that each wave of novel media technologies advanced the cause of nonfilm music, only to be overwhelmed by the Bollywood music juggernaut. However, unlike the predigitalization wave of nonfilm music, the current age of platformization affords a more sustainable and diverse nonfilm music sector.

NONFILM MUSIC BEFORE DIGITALIZATION

The earliest Indian films were cinematic renditions of popular theater musicals. In his seminal work, Peter Manuel analyzed various nonfilm music genres, such as folk, devotional, classical, and ghazal, before and during the rise of cassette culture in India in the 1980s.[9] He explained that Bollywood songs' "modes and melodies were akin to those of [Indian] folk or light-classical music," featuring a distinctive South Asian vocal style.[10] These songs stood in stark contrast to the diversity of regional folk music that Bollywood music came to appropriate, modify, homogenize, standardize, and dominate.[11] However, the substantially cheaper production costs of cassettes, as compared with more capital-intensive gramophone records, gave an impetus to struggling nonfilm music producers and spawned a cottage industry of small record companies catering to several regional markets and fringe genres.[12] This democratizing effect of technology on recorded music was reflected in the fact that the 90 percent market share that film music enjoyed was almost halved upon the arrival of cassette technologies in India.[13] However, by the 1990s, Bollywood music was back to garnering 70 to 80 percent of record sales in India and had reasserted its dominance over Indian music culture.[14]

While the aforementioned nonfilm genres experienced a revival under the aegis of cassette technologies, the popularity of these genres never reached the heights attained in the 1990s by nonfilm Hindi pop music, colloquially known as Indipop—a portmanteau of *Indian* and *pop*, influenced musically by Western disco.[15] The eminence of Indipop was exemplified by the success of the song "Made in India" and its eponymous album, which became the first Indipop album to match the success of Bollywood music albums.[16]

In 1991, the Indian economy opened to liberalization and foreign trade, which had far-reaching effects on several aspects of quotidian life in India.[17] Peter Kvetko argued that Indipop was a manifestation of the ongoing influence of globalization that was accentuated and accelerated by successive economic liberalization policies after 1991.[18] The year 1991 also marked the launch of transnational satellite television channels in India,[19] and two international music television channels—MTV India and Channel V—soon became popular, providing the Indian version of "the establishment of music video as an integral part of the pop process."[20]

The mass popularity of Bollywood music led the government to reverse its ban on the radio broadcast of film music.[21] A similar dynamic unfolded with transnational music television channels. Initially, MTV India relied on Western music videos, while Channel V mixed Western with Indipop music videos as well, to great success, including with advertisers targeting young, urban audiences.[22] MTV India relaunched in 1996 with a new brand identity,[23] dedicating 70 percent of its programming to Bollywood music.[24] MTV India actively presented itself to Bollywood film producers and directors as the most effective television channel for promoting their films.[25] Subsequently, MTV India raced ahead of Channel V.[26] Soon enough, on both channels, Bollywood music videos featuring bona fide film stars crowded out Indipop music videos featuring fledgling singers, thereby delivering a significant blow to the democratization and diversity of Indian music. Newly launched, private FM radio stations also followed suit.[27]

From a commercial perspective, music videos served as promotional advertisements for their respective music albums.[28] With the state-owned national radio already dominated by Bollywood music, the loss of support for Indipop from music television channels severely affected the sale of Indipop albums. This is reflected in the research of Kvetko, who found that the primary economic concern of the Indipop musicians and industry executives he interviewed for his work on Indipop, from October 2000 to October 2001, was the lack of support for Indipop from music television channels, rather than the launch of Napster or rising digital music piracy, which concerned Western music industry figures (although Indipop was also affected by digital "piracy").[29]

The rise of Indipop led to the co-opting of Indipop musicians—predominantly singers—by Bollywood.[30] This practice was economically motivated and legally enabled by India's unique copyright laws. An Indian Supreme Court ruling from 1977 vested the entire first copyright of the film soundtrack in the film producer or employer who commissioned the work, unless a contract to the contrary existed.[31] Consequently, musicians' rights were nearly always transferred to film producers, with the musicians receiving one-off buyout payments. Bollywood film producers, who could now sell music rights to record companies at ever-increasing prices, began to use Indipop singers rising in popularity, thanks to generous support from music television channels. This helped shift the aesthetics of Bollywood music away from Indian folk sounds of Bollywood and toward the disco-inspired sonic aesthetic of Indipop. However, analyzing the Indian music industries a few years later, Gregory Booth noted that rather than affecting a paradigm shift in Bollywood music, the "absorption" of Indipop singers into Bollywood songs ensured the longevity of the latter and diluted the "stylistic and ideological distinctions between film-song and pop-song," while maintaining the dominance of Bollywood music.[32]

For Kvetko, Indipop singers experienced a loss of "true self-expression . . . [and] individual identity" in serving the commercial interest of films through playback singing.[33] But their co-optation for Bollywood playback singing also arguably validated their talent and turned them into stars, perhaps compensating in some cases for the alienation of their creative freedom in "a music industry designing a standardized product for a mass audience."[34]

The unfair contractual practices involved in the assignment of rights to film producers led several musicians to petition the government for protection from exploitation by record companies, advocating for musicians' right to royalties to be made unassignable—though without success.[35] Through legal machinations and boardroom politics, Indian record companies came to acquire all copyright in the underlying works of the sound recordings owned by them, frustrating the endeavors of musicians to establish fair and equitable rights.[36] Live shows became the main source of livelihood for musicians. However, as with cinema tickets and recorded music albums, one of the key factors driving demand for live shows was the visibility afforded by promotional media such as radio and music television channels. And with Indipop steadily losing ground to Bollywood music on music television channels, the degree of agency available to Indipop musicians in choosing nonfilm work over Bollywood playback singing work was limited.

Thus, Indipop musicians withdrew from Indipop. When rampant digital music piracy brought the entire Indian recorded music industry to its knees between 2003 and 2007, Indipop more or less disappeared.[37]

NONFILM MUSIC UNDER PLATFORMIZATION

The Indian music streaming economy presents a range of conundrums. The market appears to be booming, nonfilm music thriving, legal infrastructure improving, and the diversity of music expanding. However, a deeper investigation illuminates the pressures and contradictions running amok in the Indian music streaming economy.

The headline of a recent article by global music industry analyst Paulina Pchelin claimed, "India's Music Market: Only 2nd to US and Surging."[38] Another global music industry report affirmed that Indians spend 24.4 hours per week listening to music, higher than the global average of 20.7 hours.[39] However, a close reading of the article by Pchelin reveals that the rankings refer to the total volume of on-demand audio and video streams.[40] At US$0.16, the country's per capita revenue from recorded music is actually among the lowest in the world; and despite entering the top five user markets for Spotify globally in just four years, India was not a top-five revenue market for Spotify.[41] An Indian music industry report ranked India fourteenth in terms of recorded music and twenty-third in terms of music publishing revenues.[42]

An enormous value gap exists within the Indian recorded music industry. One of the important reasons for this gap is the unwillingness of Indian consumers to pay for music.[43] While the number of Indian music listeners willing to pay for premium streaming services is growing,[44] the base for this growth is merely 10.5 million, which is just a little over 0.7 percent of the Indian population.[45] Instead, the Indian music streaming market is dominated by advertising revenues, which comprise approximately 77 percent of digital music revenues.[46] Nearly half of music streaming consumption in India occurs on YouTube's "free service," reflecting not only Indian consumers' tolerance for advertisements and their unwillingness to pay for subscriptions but also their preference for engaging with music visually—a legacy of the deep-rooted association of film and music.[47]

Among the audio MSPs, Spotify leads the Indian market with a 26 percent share of music streams.[48] Its biggest competitor is a local MSP, JioSaavn, owned by the largest telecom company in India, which has surpassed Spotify in monthly active users by bundling its music streaming service with the telecom services of its parent company.[49] Therefore, despite Spotify leading in terms of music streams, the market is led by a local platform in terms of users. As in so many other markets discussed in this book, other significant MSPs in the market include YouTube Music, Apple Music, and Amazon Music. But Spotify is the only MSP in the country that is not owned by a company with diversified business interests. For all the other MSPs, music is used by their parent companies to attract customers in order to sell other products and services to them.[50] However, with free, ad-supported music available on YouTube and the radio, MSPs have been unable to achieve profitability in the low-paying Indian market and are working to drive consumers toward the more profitable subscription model.[51] For example, Gaana, a local MSP, was once the largest MSP in India but was compelled to switch to an entirely subscription-based model to survive, forfeiting substantial market share in the process.[52] In contrast, Resso, an international MSP, shut down its operations in India.[53]

The complex dynamics of the Indian music streaming economy aside, the high volume of music streams in India has helped to raise the profile of Indian musicians on the global scale. For example, Indian nonfilm musician King broke into the global top song and album charts on Spotify;[54] Indian film musician Arijit Singh outranked Taylor Swift and BTS among the top ten most followed musicians on Spotify;[55] Indian-Punjabi musician Diljit Dosanjh performed at Coachella;[56] Indian metal band Bloodywood featured on the Billboard and UK charts;[57] and seven out of the top ten musicians on YouTube's global chart in early 2023 were from India.[58]

These examples illuminate some interesting undercurrents. First, except for Bloodywood, all the musicians named above have been supported by the major domestic (T-Series, Saregama, and Zee Music) or international (Universal Music, Sony Music, and Warner Music) record companies.[59] Second, Diljit Dosanjh (Punjabi) and two of the Indian musicians (Bhojpuri) in YouTube's top-ten global music chart represent regional Indian languages. Third, Arijit Singh has recently forayed into nonfilm music but is predominantly a Bollywood playback singer, as are the other five Indian musicians in YouTube's top ten global music chart. Fourth, the two Bhojpuri musicians in YouTube's top-ten global music chart work on both nonfilm and film music projects; Diljit Dosanjh has acted in and sung for Bollywood and Punjabi films; and King has also alluded to "major upcoming Bollywood features."[60] Fifth, for all seven Indian musicians in YouTube's top ten global music chart, most of their streams come from India, and the same is expected to hold true for most other Indian musicians as well.[61]

Thus, we find that Indian musicians can pursue both nonfilm and film music careers in the music streaming economy, albeit with support from the major record companies. The above examples also reflect the wider trend that, against the global average of 49 percent, 71 percent of music listening time in India is devoted to domestic music.[62] Even on Western MSPs, domestic music is resisting usurpation by dominant Western cultural flows, as theorized by platform imperialism,[63] while regional music is challenging the domestic cultural hegemony of Hindi music. In 2022, of Spotify India's top ten most-streamed songs, four were in Punjabi and one in Tamil; on Apple Music India, Punjabi hits comprised eight of the top ten songs.[64] Besides increasing the representation of regional music, the Indian platform economy has also engendered local underground genres such as rap and hip-hop, which, particularly in their contemporary form, are growing in the nonfilm sector and have also been appropriated by mainstream Bollywood after starting to achieve mass popularity.[65] Given the strong audience affinity for domestic music, it is not surprising that platforms such as Spotify and YouTube have indigenized their products and regionalized their marketing campaigns in India.[66] For example, YouTube changed its numbering format to display video views to Indian users in the more familiar local nomenclature of lakhs and crores[67] rather than millions and billions.[68]

Nonetheless, such is the dominance of Hindi that it enjoys a 64 percent share of all digital music consumption in India.[69] The share of film music has declined from approximately 80 percent four years ago to 63 percent.[70] Nevertheless, Bollywood music accounts for 60 percent of all digital music consumption.[71] As a result, some argue that, in a bid to amass subscribers, MSPs, both editorially and algorithmically, tend to reduce music diversity by reinforcing preexisting preferences for Bollywood music.[72]

The limited ability of MSPs to significantly alter the shape of the Indian recorded music industry is especially highlighted when we consider independent music—that segment of nonfilm music that is not supported by domestic or international majors. While a Federation of Indian Chambers of Commerce and Industry and Ernst and Young India report[73] pegged nonfilm music at a 37 percent share of digital listenership, it did not further bifurcate this share into independent and major record company–supported music. Considering that YouTube commands the lion's share of music consumption in India, as discussed above, one might expect independent music to thrive on the platform, which encourages user-generated content. However, two of the top ten most-subscribed Indian channels on YouTube are operated by two major domestic record companies (T-Series and Zee Music),[74] while the only other two music channels on the list are operated by traditional vertically integrated film production companies (Tips Industries and Shemaroo Entertainment).[75] Therefore, the majority share of revenue from the advertising-driven Indian music streaming economy is still enjoyed by major record companies and long-established film producers who own the rights to their hit soundtracks, with limited revenue going to vast number of musicians operating as independents.

However, hope for independent musicians lingers in social media, which, along with YouTube, is the primary source of music discovery for Indian listeners.[76] Independent musicians are also encouraged by stories such as those of King, who, as discussed above, acquired a fan following independently on YouTube before signing with Warner Music India.[77] His story echoes that of the Canadian singer Justin Bieber, who was discovered on YouTube by his future manager and then signed to a major record company, which expanded his reach and paved his way to celebrity.[78] Arguably, the path from discovery on YouTube to global stardom is complex, with such cases being rarer in the relatively smaller Indian music market. Therefore, as scholars such as David Hesmondhalgh caution, musicians ought to be wary of confusing these exceptional instances of upward mobility with a false sense of democratization.[79] Nonetheless, the deluge of short-format video platforms that emerged to fill the void left by the Indian government's ban on TikTok has revitalized hopes for music discovery.[80] One report estimates that nearly one-fifth of overall music listening time per week takes place on these platforms, second only to YouTube.[81] But the influence of Bollywood and major record companies dominates social media and short-video-format platforms as well. For example, Neha Kakkar, a popular Indian singer with an enviable catalog of Bollywood and nonfilm music, supported by the domestic and international majors, has attracted seventy-eight million followers, making her the most-followed Indian musician on Instagram.[82]

Apart from a sense of false hope which tends to equate music distribution with discovery on the demand side, on the supply side, the cost of digital music production has been substantially reduced by software, as has the cost of music distribution to MSPs via digital music distribution platforms engendered globally by the music streaming economy.[83] This has greatly reduced the entry barrier for independent, part-time, and hobbyist musicians (including this author) to try their luck by publishing music to MSPs themselves.[84] However, with multiple songs, these costs add up over time and increase the financial risk of production for those musicians hoping to catch the attention of audiences, record companies, and film producers. In addition to self-publishing on user-generated content platforms discussed above, these musicians, and more recently even film producers,[85] distribute their music to global MSPs in exchange for paying a commission to distribution platforms on the revenues earned from their music streams, plus a fee for any additional value-added services procured from these distributors.[86] Crucially, these distribution platforms allow musicians to publish their music globally by bypassing record companies and without surrendering their copyright.[87]

However, despite these novel business models, long-standing legal battles over the country's copyright laws present a further barrier to the sustenance of all Indian musicians, but especially independent musicians. Through a concerted effort to correct predigitalization copyright issues, some of which were highlighted in the previous section, the Copyright Amendment Act was passed in 2012. A key inclusion in this legislation was the inalienable right to royalties for lyricists and composers of musical works—that is, these musicians could transfer their copyrights for exploitation, but under no circumstances could their right to royalty be annulled; any contract to the contrary was deemed void.[88] But with all music rights traditionally bundled and transferred from film producers to record companies, the music publishing industry in India is still nascent, and divergent court interpretations of the amendments have led to low awareness and low compliance from the end users of music, resulting in multiple litigations.[89] On the other hand, fledgling independent musicians seeking deals from film producers and record companies are still pressured to forsake their right to royalties in exchange for one-time buyout payments, lack education about their legal rights, and have little bargaining power to oppose exploitative practices for fear of losing future work.[90] Dismally, according to one report, "only 13,500+ of an estimated 60,000+ music creators [in India] have registered with their copyright society."[91] Furthermore, the Copyright Amendment Act also recognized royalties for performers, but this, too, was contested; and an agreement was only recently struck between the collection society for performers and the trade organizations representing the record companies for royalties to be paid out to Indian performers.[92] Given the still maturing legal infrastructure and the delicate music streaming economy struggling to achieve profitability, one can expect meager incomes accruing to musicians, especially independent musicians, from their recorded music when compared with live shows, which remain the staple for most musicians even in the platform economy— a fact that was painfully highlighted during the COVID-19 pandemic.[93]

Finally, another area that suffered huge losses during the pandemic but is often ignored in industry reports is India's huge informal music sector. Estimated to be between seventy-four and three hundred times the size of the formal music industry, the informal music sector includes DJs, brass band members, sound engineers, teachers, and manufacturers. Most in this sector earn a living from offline or seasonal work and lack social security in the face of calamities such as the pandemic.[94] With a little over fourteen million people, the Indian informal music sector employs more people than other major Indian employers, such as the railways, the government, the telecom industry, and the textile industry.[95] However, while the Indian government has developed focused policies for growing the telecom, textile, and railway sectors, the Indian music industries have largely been bereft of state attention.[96] Dismally, workers in the Indian informal music sector earn less than the "median salary of an unskilled worker" in India.[97] Bringing the workers of the Indian informal music sector into the formal music economy thus represents not only an immense potential for the Indian music industry but also the degree of reform needed through government policies.

CHANGES AND CONTINUITIES

Having analyzed the Indian recorded music industry in the pre- and postdigitalization eras, we are now well poised to examine the changes and continuities in the industry and assess the role of platformization.

At each stage of technological development, we find a certain democratization of music culture in India. While cassettes democratized music audio, transnational music television channels democratized music videos, and digital platforms have democratized a diverse range of audiovisual musical content. However, at each stage, we also consistently find the dominating and delimiting influence of Bollywood music. The cases of radio and transnational music television channels in the predigitalization era, and global MSPs in the current era of platformization, highlight how these media have been compelled to fulfill and reinforce the audience demand for Bollywood music, at the expense of the nonfilm music sector.

Deeply linked to the tenacious cultural hegemony of Bollywood music is the persistent oligopolistic control wielded by the major, especially domestic, record companies, which have built deep catalogs of Bollywood music over the years and wield significant bargaining power over the MSPs. For example, in 2019, the record company Saregama obtained an injunction from an Indian high court to take down 120,000 songs from Spotify just two months after the platform launched in India,[98] and Spotify had to comply until it reached a new agreement with Saregama a year later.[99] Furthermore, while Indipop musicians were unable to distribute their music in physical retail or promote their music on radio and television without the support of record companies, even today, the purportedly democratized self-publishing opportunities for independent musicians are thwarted by their inability to match the aggressive marketing tactics and strong professional networks of record companies.[100] If anything, the demise of Indipop—accelerated by the loss of investment from record companies following the withdrawal of promotional support from broadcast media—should apprise budding musicians of the ephemerality of musical careers and the dominant role of record companies in shaping the music culture of India.

It is also evident from the analysis that the Indian music market does not present a clear example of domination by Western cultural flows. The cases of both the transnational music television channels and the MSPs demonstrate how they have indigenized their businesses to survive in the resilient domestic cultural market.[101] As Christine Ithurbide highlights, the strong affinity of Indians for Indian music has negated the need for any "minimum quotas for domestic music on radio as is observed in some other countries."[102] International music accounts for a negligible share of all digital music streams in India;[103] and while its consumption is increasing in absolute terms, its share is diminishing.[104] On the other hand, the Indian music market has witnessed an increase in the acquisition cost of regional music,[105] as well as the acquisition of regional record companies by national ones,[106] particularly in the South Indian languages of Telugu and Tamil.

Another market dynamic that has extended into the current era of platformization is the ability of Bollywood to operate as a "large-scale sniffing machine"[107] that co-opts alternative music cultures on the brink of mass popularity and then amplifies them nationally. However, the crucial change is that, while Bollywood's appropriation led to the desertion of Indipop by its musicians—who had lost institutional support for their music—nonfilm musicians in the platform era can advance their careers by engaging in both film and nonfilm music projects. This shift has been driven by both a growing acceptance for nonfilm music in India and the self-publishing affordances of platformization. Since 1991, liberalization policies of successive governments have fueled the growth of the Indian middle class and engendered sociocultural changes across various aspects of quotidian life, including films, fashion, food, travel, and music.[108] As the burgeoning middle class has been increasingly exposed to international nonfilm music through expanding international connections and strengthening diasporic ties[109] over the past three decades, its acceptance of local nonfilm music has also increased. Supporting this latent demand, the self-publishing affordances of platformization that allow nonfilm musicians to grow and maintain audiences directly, independent of walled-garden media such as films, radio, and television, seem to have delivered a positive impact on the diversity and sustainability of a parallel nonfilm recorded music industry. Consequently, leading Indian hip-hop musicians, such as Badshah, Raftaar, and Yo Yo Honey Singh—who have written, composed, and sung for Bollywood songs—have continued to pursue their nonfilm careers through their YouTube, social media, and record company engagements.[110] Furthermore, as discussed above in the case of Arijit Singh, film musicians today are also pursuing nonfilm music careers and, along with their nonfilm peers, many have launched their own record companies.[111] Therefore, the lines between film and nonfilm musicians are rapidly blurring, and musicians are simultaneously exploiting all three channels—film, record company, and independent—for distributing and marketing their music.

Nonetheless, as noted in the previous section, the concentration of power among a few major record companies, weak copyright laws, and the dominance of Bollywood music leaves little scope for earning a sustainable livelihood for most independent musicians. While copyright laws have been substantially and significantly strengthened to support musicians, the benefits of these changes have not percolated down to most musicians. Revenues for record companies grew by 17 percent in 2023,[112] but the amounts received by most artists are negligible.[113] Unable to survive on the low incomes from MSPs and advertising-driven platforms,[114] most musicians in the platform era must depend, like those in the predigitalization era before them, on multiple sources of income outside the sales of recorded music, especially live shows.

CONCLUSION

This chapter investigated the impact of music streaming on the Indian recorded music industry by analyzing two distinct waves of nonfilm music in the pre- and postdigitalization eras. The analysis found that long-standing issues from the predigitalization era—such as the cultural hegemony of Bollywood music, the oligopoly of major domestic record companies, and an inefficient legal system—extend into the current era, and platformization has done little to alleviate these problems. On the other hand, there is a marked increase in the diversity of music, especially with the rise of regional and nonfilm music. Furthermore, from being the only sustainable career option in the predigitalization era, Bollywood music is now one of several options available to musicians keen on pursuing a long-term career in recorded music. However, these options also seem to have created a sense of false hope among independent musicians—a category that was practically nonexistent in the predigitalization era—who persistently risk investing their own finances in music production, publishing, and marketing with the aim of attracting the attention of listeners and the powers that be. Only time will tell how many of these hopes can be fulfilled by the apparent democratizing effect of platformization. Until then, live shows have been, are still, and seem likely to remain, for the near future, the basis of musical income for most musicians in India's music streaming platform economy.

NOTES

1. The word *Bollywood* is sometimes misleadingly used to refer to the Indian film industry as a whole, whereas Bollywood is the colloquial name given to the Hindi film industry based in Bombay (now Mumbai).

2. Aswin Punathambekar, *From Bombay to Bollywood: The Making of a Global Media Industry* (New York: NYU Press, 2013).

3. While nonfilm music can also imply international music, in this chapter, it refers strictly to Indian nonfilm music.

4. "Spotify Highlights Growth of Independent Music in India," *Music Plus*, March 12, 2024, www.musicplus.in/spotify-highlights-growth-of-independent-music-in-india/.

5. Lata Jha, "Labels, Artistes Tap Growing Demand for Non-Film Music," *Mint*, April 20, 2022, www.livemint.com/industry/media/labels-artistes-tap-into-growing-demand-for-non-film-music -11650428801812.html.

6. "Spotify Highlights Growth of Independent Music in India."

7. International Federation of the Phonographic Industry (IFPI), *Engaging with Music* (London: IFPI, 2023), 25.

8. "Spotify Highlights Growth of Independent Music in India."

9. Peter Manuel, *Cassette Culture: Popular Music and Technology in North India* (Chicago: University of Chicago Press, 1993).

10. Manuel, *Cassette Culture*, 50.

11. Manuel, *Cassette Culture*, 52–54.

12. Manuel, *Cassette Culture*, 64.

13. Peter Manuel, "The Cassette Industry and Popular Music in North India," *Popular Music* 10, no. 2 (1991): 189–91.

14. Jocelyn Cullity, "The Global Desi: Cultural Nationalism on MTV India," *Journal of Communication Inquiry* 26, no. 4 (2002): 413–14.

15. Peter Kvetko, "Can the Indian Tune Go Global?," *TDR / The Drama Review* 48, no. 4 (2004): 185.

16. Ulrike Rohn, *Cultural Barriers to the Success of Foreign Media Content: Western Media in China, India, and Japan* (Frankfurt: Peter Lang GmbH, 2010), 170.

17. Asha Kasbekar, *Pop Culture India! Media, Arts, and Lifestyle* (Santa Barbara, CA: ABC-CLIO, 2006).

18. Peter Kvetko, "Indipop: Producing Global Sounds and Local Meanings in Bombay" (PhD diss., University of Texas at Austin, 2005).

19. Kasbekar, *Pop Culture India!*, 7.

20. Keith Negus, *Producing Pop: Culture and Conflict in the Popular Music Industry* (London: Edward Arnold, 1992), 93.

21. Manuel, *Cassette Culture*, 39–40.

22. Kasbekar, *Pop Culture India!*, 21–22.

23. Punathambekar, *From Bombay to Bollywood*, 96.

24. Cullity, "Global Desi," 414.

25. Punathambekar, *From Bombay to Bollywood*, 103–4.

26. Cullity, "Global Desi," 413.

27. Gregory D. Booth, "Preliminary Thoughts on Hindi Popular Music and Film Production: India's 'Culture Industry(ies)', 1970–2000," *South Asian Popular Culture* 9, no. 2 (2011): 217; "Radio, TV Didn't Encourage Indipop: Lesle Lewis," *Business Line*, December 23, 2012, www.thehindubusinessline .com/news/variety/Radio-TV-didn%E2%80%99t-encourage-Indipop-Lesle-Lewis/article20543648.ece.

28. Jayson Beaster-Jones, "A. R. Rahman and the Aesthetic Transformation of Indian Film Scores," *South Asian Popular Culture* 15, no. 2–3 (2017): 440.

29. Peter Kvetko, "It's Rocking? Exploring Sound and Intimacy through Mumbai's Faltering Indipop Music Industry," in *Indian Sound Cultures, Indian Sound Citizenship*, ed. Laura Brueck, Jacob Smith, and Neil Verma (Ann Arbor: University of Michigan Press, 2020), 74.

30. Kvetko, "Indipop," 288.

31. Prashant T. Reddy, "The Background Score to the Copyright (Amendment) Act, 2012," *NUJS Law Review* 5, no. 4 (2012): 484.

32. Booth, "Preliminary Thoughts on Hindi Popular Music and Film Production," 217.

33. Peter Kvetko, "Private Music: Individualism, Authenticity and Genre Boundaries in the Bombay Music Industry," in *Popular Culture in a Globalised India*, ed. K. Moti Gokulsing and Wimal Dissanayake (London: Routledge, 2008), 116.

34. Manuel, *Cassette Culture*, 54.

35. Aditya Lal, David Hesmondhalgh, and Charles Umney, "The Changing Shape of the Indian Recorded Music Industry in the Age of Platformisation," *Contemporary South Asia* 31, no. 2 (2023): 298.

36. Reddy, "Background Score to the Copyright (Amendment) Act." See this journal article for a more detailed history of the legal, political, economic, and ideological factors that influenced the evolution of Indian copyright laws, particularly in relation to the Indian recorded music industry.

37. Gregory Booth, "Copyright Law and the Changing Economic Value of Popular Music in India," *Ethnomusicology* 59, no. 2 (2015): 281.

38. Paulina Pchelin, "India's Music Market: Only 2nd to US and Surging," *Luminate*, December 5, 2023, https://luminatedata.com/blog/indias-music-market-only-2nd-to-us-and-surging/.

39. IFPI, *Engaging with Music*.

40. Pchelin, "India's Music Market."

41. Amit Gurbaxani, "India Is Driving Spotify's Global Growth—Will Revenue Follow?," *Billboard*, April 4, 2023, www.billboard.com/pro/india-spotify-global-growth-revenue/.

42. Ernst and Young India (EY), *The Music Creator Economy: The Rise of Music Publishing in India* (New Delhi: EY, 2023), 14.

43. Mansi Kedia, Aarti Reddy, Richa Sekhani, and Saptorshi Gupta, *The Untold Potential of India's Informal Music Industry* (New Delhi: Indian Council for Research on International Economic Relations, 2022), 1.

44. Pchelin, "India's Music Market."

45. Federation of Indian Chambers of Commerce and Industry (FICCI) and EY, *Shape the Future* (New Delhi: EY, 2025), 55.

46. FICCI and EY, *#Reinvent* (New Delhi: EY, 2024), 170.

47. FICCI and EY, *Windows of Opportunity* (New Delhi: EY, 2023), 81.

48. Sohini Mitter, "Spotify Leads JioSaavn, YouTube Music, Others with 26% Share in India's Audio-Streaming Market," *Business Today*, April 21, 2023, www.businesstoday.in/trending/entertainment/story/spotify-leads-jiosaavn-youtube-music-others-with-26-share-in-indias-audio-streaming-market-378305-2023-04-21.

49. Gurbaxani, "India Is Driving Spotify's Global Growth."

50. Christine Ithurbide, "Telecom and Technology Actors Repositioning Music Streaming," in *Platform Capitalism in India: Global Transformations in Media and Communication Research*, ed. Adrian Athique and Vibodh Parthasarathi (Cham, Switzerland: Palgrave Macmillan, 2020), 115.

51. FICCI and EY, *Windows of Opportunity*, 222.

52. FICCI and EY, *Windows of Opportunity*, 222.

53. Murray Stassen, "TikTok Parent to Shut Down Music Streaming Service Resso in India," *Music Business Worldwide*, January 11, 2024, www.musicbusinessworldwide.com/tiktok-parent-to-shut-down-music-streaming-service-resso-in-india/.

54. "Indian Artist 'King' Rules Spotify Global Charts," *Loudest*, November 25, 2022, https://loudest.in/news/indian-artist-king-rules-spotify-global-charts-15996.html.

55. "Arijit Singh Creates History Beating Taylor Swift Crossing 59 Million Followers, Almost Near to Cross Eminem in the List of Most Followed Artists on Spotify," *Koimoi*, accessed February 1, 2023, www.koimoi.com/bollywood-news/arijit-singh-creates-history-beating-taylor-swift-crossing-59-million-followers-almost-near-to-cross-eminem-in-the-list-of-most-followed-artists-on-spotify/.

56. Zoya Mateen, "Diljit Dosanjh: From Coachella to Fallon, Decoding His Stardom," *BBC News*, June 28, 2024, www.bbc.com/news/articles/cxo2x58voe5o.

57. Anurag Tagat, "Bloodywood Make Indian Metal History with Debut Album 'Rakshak,'" *Rolling Stone India*, March 4, 2022, https://rollingstoneindia.com/bloodywood-rakshak-album-billboard-charts/.

58. Stuart Dredge, "Indian Artists Are Still Dominating YouTube's Global Music Chart," *Music Ally*, January 30, 2023, https://musically.com/2023/01/30/indian-artists-are-still-dominating-youtubes-global-music-chart/.

59. The recorded music industry in India, unlike in most other parts of the world, is dominated by domestic major record companies rather than international ones.

60. "King," Warner Music India, accessed July 2, 2024, https://warnermusicindia.co/artists/king.

61. Dredge, "Indian Artists."

62. Indian Music Industry (IMI), *Music: A Savior in COVID-19 Times* (Mumbai: IMI, 2022), 21.

63. Dal Yong Jin, *Digital Platforms, Imperialism and Political Culture* (New York: Routledge, 2015), 119.

64. Amit Gurbaxani, "India," in *The Report: Knowledge for the Global Music Business* (London: Music Ally, 2023), 51–52.

65. Elliot Cardozo, "Hip Hop Goes to B-Town: Bollywood's Assimilation of the Underground Aesthetic," *SRFTI Take One* 2, no. 1 (2021): 26–43.

66. "Spotify India Turns Three, Focus Remains on Growing the Market," *BestMediaInfo*, February 25, 2022, https://bestmediainfo.com/2022/02/spotify-india-turns-three-focus-remains-on-growing-the-market.

67. One lakh equals hundred thousand. One crore equals ten million.

68. Ashri Khandelwal, "YouTube Now Showing View Counts in Lakhs and Crores Instead of Millions and Billions for Some Indian Users," *Times Now News*, last updated April 18, 2020, www.timesnownews.com/technology-science/article/youtube-now-showing-view-counts-in-lakhs-and-crores-instead-of-millions-and-billions-for-some-indian-users/579814.

69. FICCI and EY, *Shape the Future*, 187.

70. FICCI and EY, *Shape the Future*, 183.

71. FICCI and EY, *Shape the Future*, 183.

72. Ithurbide, "Telecom and Technology Actors," 122.

73. FICCI and EY, *Shape the Future*, 183.

74. Since 2019, the Bollywood music channel operated by T-Series was the most-subscribed YouTube channel in the world, dropping to second place in June 2024. See Liv McMahon, "MrBeast: YouTuber Topples T-Series for Most Subscribers," *BBC News*, accessed July 1, 2024, www.bbc.com/news/articles/c8440yxd92xo.

75. FICCI and EY, *Shape the Future*, 183.

76. FICCI and EY, *Windows of Opportunity*, 81.

77. "Warner Music India Signs Indian Breakout Artist King," *PR Newswire*, December 15, 2022, www.prnewswire.com/news-releases/warner-music-india-signs-indian-breakout-artist-king-301703627.html; Punita Sabharwal, "Reign of the 'King': Arpan Kumar Chandel," *Entrepreneur*, February 9, 2024, www.entrepreneur.com/en-in/lifestyle/reign-of-the-king-arpan-kumar-chandel/469445.

78. David Hesmondhalgh, "Have Digital Communication Technologies Democratized the Media Industries?," in *Media and Society*, ed. James Curran and David Hesmondhalgh, 6th ed. (London: Bloomsbury, 2019), 116.

79. Hesmondhalgh, "Digital Communication Technologies," 116.

80. Before it was banned, TikTok had become an important platform for the discovery of independent music in India, with six out of the top ten Indian songs on the platform belonging to the independent stable. See Aranyaka Verma, "India's TikTok Ban a Blow to Music Discovery and Independent Artists," *Outlook*, last updated July 19, 2020, www.outlookindia.com/website/story/opinion-indias-tiktok-ban-a-blow-to-music-discovery-and-independent-artists/356991. A strong correlation between the virality of a song on TikTok and its listenership on Spotify had also been established.

See Gaurav Wadhwa, "How Short Video Apps Are Creating a New Generation of Indian Pop Stars," *ET Brand Equity*, last updated December 13, 2022, https://brandequity.economictimes.indiatimes.com/news/digital/how-short-video-apps-are-creating-a-new-generation-of-indian-pop-stars/96180602.

81. IMI, *Music*, 11.

82. "Top 10 Indians with Highest Followers on Instagram in 2024," *Forbes India*, January 24, 2025, www.forbesindia.com/article/explainers/indians-highest-followers-instagram/85725/1.

83. Kedia, Reddy, Sekhani, and Gupta, *India's Informal Music Industry*, 13.

84. Sometimes, these musicians are problematically referred to as do-it-yourself (DIY) musicians. See Shuwen Qu, David Hesmondhalgh, and Jian Xiao, "Music Streaming Platforms and Self-Releasing Musicians: The Case of China," *Information, Communication and Society* 26, no. 4 (2021): 699–715.

85. "Believe Enters the Bollywood OST Market with Exclusive Distribution Deal," *Music Plus*, April 4, 2023, www.musicplus.in/believe-enters-the-bollywood-ost-market-with-exclusive-distribution-deal/.

86. "The Mechanics of Distribution: How It Works, Types of Music Distribution Companies and 35 Top Distributors" *Soundcharts*, December 31, 2023, https://soundcharts.com/blog/music-distribution.

87. The digital music distribution ecosystem in India is dominated by the local offices of global distributors such as Believe, CD Baby, TuneCore (owned by Believe), DistroKid, the Orchard (owned by Sony Music), and Horus Music, to name a few. Recently, one of the leading Indian distributors, OKListen, was acquired by Sony Music and relaunched as the local arm of the Sony Music subsidiary AWAL. See Daniel Tencer, "AWAL Expands into India, South Asia with OKListen Acquisition," *Music Business Worldwide*, June 26, 2023, www.musicbusinessworldwide.com/awal-expands-into-india-south-asia-with-oklisten-acquisition/. Apart from these distributors, multichannel networks (MCNs) such as One Digital Entertainment focus on managing the YouTube channels for various smaller record companies and celebrities, including musicians. See Jai Vardhan, "The Startups behind the Ultimate Rise of Multi-Channel Networks (MCN)," *YourStory*, February 16, 2015, https://yourstory.com/2015/02/rise-of-multi-channel-networks-mcn.

88. Prashant T. Reddy and Sumathi Chandrashekaran, *Create, Copy, Disrupt: India's Intellectual Property Dilemmas* (New Delhi: Oxford University Press, 2017), 197.

89. EY, *Music Creator Economy*, 11.

90. Ithurbide, "Telecom and Technology Actors," 123.

91. EY, *Music Creator Economy*, 28.

92. Andre Paine, "PPL Signs Landmark Indian Deal with ISAMRA," *Music Week*, January 23, 2024, www.musicweek.com/labels/read/ppl-signs-landmark-indian-deal-with-isamra/089152.

93. Taniya Roy, "India's Indie Musicians Struggle as COVID Restrictions Reduce Their Incomes by Half," *The Wire*, August 15, 2021, https://thewire.in/culture/indias-indie-musicians-struggle-as-covid-restrictions-reduce-their-incomes-by-half.

94. Kedia, Reddy, Sekhani, and Gupta, *India's Informal Music Industry*.

95. Kedia, Reddy, Sekhani, and Gupta, *India's Informal Music Industry*, 4.

96. Kedia, Reddy, Sekhani, and Gupta, *India's Informal Music Industry*, 4.

97. Kedia, Reddy, Sekhani, and Gupta, *India's Informal Music Industry*, 6.

98. Daniel Sanchez, "Indian High Court Orders Spotify to Delete 120,000 Songs," *Digital Music News*, April 24, 2019, www.digitalmusicnews.com/2019/04/24/spotify-v-saregama-india/.

99. Aman Rawat, "Spotify Signs Licensing Deal with Saregama for India," *Inc42*, accessed May 11, 2020, https://inc42.com/buzz/spotify-signs-licensing-deal-with-saregama-for-india/.

100. Kedia, Reddy, Sekhani, and Gupta, *India's Informal Music Industry*, 13.

101. Eventually, both MTV India and Channel V migrated toward local youth-oriented reality-TV programming. Today, Channel V is no longer in operation, whereas another television channel, MTV Beats, has been spun off to continue broadcasting Indian music videos.

102. Ithurbide, "Telecom and Technology Actors," 121.

103. FICCI and EY, *Shape the Future*, 187.

104. Jacca, "The State of Indian Artists and Music Streaming in 2021," *RouteNote*, October 29, 2021, https://routenote.com/blog/the-state-of-indian-artists-and-music-streaming-in-2021/.

105. FICCI and EY, *Shape the Future*, 182.

106. FICCI and EY, *Windows of Opportunity*, 220.

107. Gautam Pemmaraju, "Hindi Film Music Is a Large-Scale Sniffing Machine: COO Times Music," *Outlook*, January 3, 2023, www.outlookindia.com/art-entertainment/hindi-film-music-is-a-large-scale-sniffing-machine-coo-times-music-magazine-249253.

108. Kasbekar, *Pop Culture India!*

109. Manuel, *Cassette Culture*, 144.

110. Cardozo, "Hip Hop Goes to B-Town," 32.

111. Samiksha Chaudhary, "Game-Changing Independent Indian Music Labels on Our Radar," *Homegrown*, July 27, 2022, https://homegrown.co.in/homegrown-creators/game-changing-independent-indian-music-labels-on-our-radar; Anurag Tagat, "New Record Labels Have Sprung Up to Claim Space in Indian Non-Film Music," *Rolling Stone India*, July 16, 2020, https://rollingstoneindia.com/new-record-labels-non-film-music-vb-merchant-tm-at-azaad/.

112. FICCI and EY, *#Reinvent*, 171.

113. Ithurbide, "Telecom and Technology Actors," 120.

114. Kedia, Reddy, Sekhani, and Gupta, *India's Informal Music Industry*, 14.

10

Independent Music Creators and Self-Releasing in China

A History of Platformization

Zhongwei (Mabu) Li and D. Bondy Valdovinos Kaye

INTRODUCTION

Taking a historical approach, this chapter investigates the self-releasing practices of independent music creators on Chinese music streaming platforms (MSPs), unpacking old and new questions of power and agency and demonstrating ways that these dynamics are conditioned by the history of music platformization in China. Much previous work studying platformization in the realm of music culture has been conducted in so-called Western countries. In this chapter, we address a major gap in research on the history of platformization (and digitalization, to a lesser extent) in the Chinese context, complementing existing approaches with a perspective focused on the working lives of music creators. Using mainland China as a case study, we extend long-standing traditions of critical research on capitalist music industries into the contemporary era of informational capitalism, adding empirical insights on how music creators make sense of and negotiate with emergent relations structured by platformization and datafication in the age of streaming.[1]

This chapter is structured as follows. In the next section, we briefly review the dynamics of power and agency that affect music creators under informational capitalism. Next, we outline three historical shifts that sketch the contours of music platformization in China: datafication, copyrightization, and infrastructural (en)closure. We then begin our empirical analysis with a brief history of independent music production and distribution in China, charting a trajectory from the pre-digital underground era to the virtual scenes empowered by digital self-releasing on the early Chinese internet. We continue our empirical analysis by examining how self-releasing practices were transformed by the arrival of a platform-based system in China and how independent music creators experienced this process of platformization. We discuss issues shared by our participants, including the datafication of artist identity, the propertization of musical work, and the predicament of alienation. We demonstrate that beneath these changing dynamics of power and agency are broader structural shifts in communication infrastructure, industrial configuration, and frameworks of governance, pushing toward an all-encompassing regime of music commodification in China's newly consolidated platform economy. We conclude with critical reflections on the relationship between independent music, technological change, and capitalism.

1   CRITICAL RESEARCH ON MUSIC CREATORS UNDER INFORMATIONAL CAPITALISM

One useful way to approach questions of power and agency for music creators is by conceptualizing the music recording industry as a structure of capitalism, which conditions the social practice of cultural production.[2] Following this approach helps us to attend to normative questions about the experiences of contemporary music creators in their historical context. For reasons of space, here we focus briefly on three key structures. The first is copyright law. While copyright provides remuneration from musical works, some music industry scholars have critiqued the fundamental assumptions that underpin copyright regimes.[3] The second structure is the recording contract. Recording contracts offer musicians opportunities for commercial rewards from their work, but at the expense of control.[4] In the streaming era, many different distribution services exist that allow music creators to circulate their work without a record contract in the traditional sense.[5] While these "self-releasing" creators may avoid some of the pitfalls of a recording contract, they are nevertheless constrained by a different set of structures on digital platforms. We follow legal scholar Julie Cohen's conceptualization of informational capitalism to examine the impact of platforms and platformization on music creators' agency. Cohen highlights the political and economic transformations that have given rise to new power relations at the intersections of society, industry, and technology and outlines the legal structures that give platforms substantial control over informational resources, such as music copyrights.[6] Now, in addition to longstanding industry actors, digital platforms enforce copyright and set contractual terms that create new restrictions for self-releasing creators. Extending critiques of capitalism, a third key structure in the platform era is data and regimes of datafication more broadly. On digital platforms, data increasingly structures relationships between creators, platforms, audiences, and music.[7] As previous scholars have noted, the social reality of music is changing as intermediaries and industry actors rely more heavily on data for strategic planning and predictive power.[8] Rather than looking at the more functional ways music creators might instrumentalize data in their everyday practice, we seek to understand how they "deal with the increasing embedding of quantification, measurement, and calculation in their everyday lives and practices" in the context of music platformization.[9]

2   MUSIC PLATFORMIZATION IN CHINA: THREE HISTORICAL SHIFTS

While the term "platformization" has been used in a general sense to describe the effects and consequences of digital platforms on various scales, we see platformization as essentially a macrohistorical process: the penetration of an overarching "platform logic" in all spheres of human life in recent history.[10] This approach calls for a platform analysis based on concrete historical contextualization: not one homogenous narrative of platformization, but many intersecting trajectories of platformization, as outcomes of historical path dependence conditioned by existing sociocultural, economic, and legal structures in local contexts. In this sense, to study platformization is not only to investigate its underlying technologies (software, protocols, physical and digital infrastructures), but also to explore historical shifts in other spheres of society embedded within the transformative mechanisms of platform logic. We identify three such historical shifts.

The first shift focuses on information infrastructures. Scholars taking a science and technology studies (STS) and media studies approach have observed a general historical trend that may be termed *infrastructural (en)closure.*[11] Looking back at the early history of the "open web," this view theorizes platformization as a process of "(en)closure," in which openness and generativity—once the fundamental principles of internet architecture—have been gradually closed off in favor of a more commercial, platform-based infrastructural system, a process characterized by the replacement of personal computers with "tethered" devices such as smartphones and the popularization of "trusted systems," which grant copyright owners an unprecedented degree of control.[12] Zhongwei (Mabu) Li and David Hesmondhalgh recount the form this infrastructural (en)closure took in the Chinese music context, from the early Chinese internet to the contemporary dominance of MSPs.[13] From the 2000s to early 2010s, the generative nature of the early internet gave rise to a coexisting diversity of both licensed and pirated digital music service models in China. As mobile devices replaced PCs to become the primary means of digital music consumption, this diversity gradually faded. Today, the few dominant MSPs in the Chinese market base all their services on a standardized, trusted-system-based model, abandoning principles of openness and generativity in favor of security and consumer convenience.

The second shift concerns the *datafication* of recorded music. From a music business studies perspective, what underlies this shift is the replacement of consumer electronics by information technology as the determining sector of the music business, pushing toward a "post-record music industry" led by "new digital conglomerates" with close ties to data industries.[14] The datafication of music denotes the process by which, in addition to other cultural and economic factors, the worth of musical commodities is increasingly determined by their value as data—that is, "content" that attracts advertising—and musicians are "redefined as content providers rather than creative producers."[15] In the Chinese context, a similar shift in which data flows and consumer attention—often referred to by the buzzword "traffic" (流量)—dictates new logics of cultural production and distribution has been noted by researchers of digital platforms.[16] Qian Zhang and Keith Negus[17] investigate how datafication has shaped the practices of cultural intermediaries in China's music industry, outlining corresponding trajectories from "music planners" at record companies to "content operators" at digital platforms.

The third shift involves *copyrightization*, or the rapid institutionalization of copyright in China. Intellectual property is a central feature in today's platformed economy of cultural production. The internationalization of Western copyright has been closely intertwined with the transnational penetration of platformization.[18] In China, the arrival of a firmly institutionalized copyright regime coincided with the embrace of new, platform-based cultural business models.[19] China did not have a legal framework for copyright protection until the 1990s.[20] Soon after, the 2010s saw an intense period of state-driven enforcement of copyright, accompanied by a decline in media piracy across the music and screen sectors,[21] argued to be the result of "international pressure, state regulation, industry self-regulation, and market competition in the post-WTO era."[22] In relation to the digital music business, this process was marked by major developments such as the third revision of the Copyright Law of China in 2012, the establishment of new IP courts in 2014, and the 2015 copyright notice issued by the National Copyright Administration of China (NCAC). The 2015 NCAC notice, which required platforms to take down all unauthorized tracks, eventually triggered a yearslong "copyright war" as MSPs, backed by Chinese tech giants, competed to bid for exclusive licensing deals with major music rights-holders. As a result, a new, formalized music streaming sector, which integrated the Chinese platform economy with the interests of local and international music industries, took form.

Together, the three shifts form a historical background against which we situate the analysis of our focus group data, to which we now turn.

3   THE PLATFORMIZATION OF DIGITAL SELF-RELEASING, AND WHAT IT MEANS TO INDEPENDENT MUSIC CREATORS

Independent Music Distribution in China:
From "Underground" to Self-Releasing

Independent music-making in the People's Republic of China, as a form of alternative cultural politics, dates back to the late 1980s. Underground rock "parties" in Beijing nourished the first generation of Chinese *yaogun* (摇滚) artists, most notably Cui Jian, who played a significant role challenging the cultural and ideological hegemony of the party-state in the 1989 democracy movement.[23] After the brief but influential "rock fad" in the early 1990s,[24] the mid-1990s saw the gradual establishment of what became the material and organizational infrastructure for a new generation of "underground" (地下) music in urban China. Much like during the heyday of alternative and indie rock and pop in the West,[25] independent labels and microlabels, rock magazines and fanzines, music bars, bookstores, and record shops became key nodes in alternative local networks of production and distribution. Thanks to the enormous flow of Western popular music recordings through "cut-outs" and piracy, what used to be a rock-dominated musical space spawned diverse musical styles with distinctive scenic politics.[26]

As in the West, digital technologies opened new doors for outsiders of official music industries in China. The early 2000s saw the mushrooming of music-themed websites, BBS forums, and file-sharing communities, which generated virtual scenes mutually imbricated with offline independent music activities.[27] As the Chinese music underground embraced digitalization, online self-releasing eventually became the default way of independent music distribution. A few companies emerged as key providers of self-releasing services online, including Neo-Cha, an early online community space for Chinese musicians and artists launched in 2006; Douban, a social networking site founded in 2005 that attracted users with its comprehensive database of information about books, films, and music releases; and Xiami Music, a peer-to-peer-based music distribution site founded in 2007. These websites quickly gathered a critical mass of musicians in the local scenes as their initial users. By the early 2010s, they had established mature self-releasing portals for a large group of unsigned music creators across China. The moniker "independent" (独立) came to be frequently attached to this group (whereas during the 1990s, the term "underground" was much preferred).[28] To a large extent, this group still carried an ideological heritage from predigital "underground" music scenes, regarding their music-making as a form of alternative, if not oppositional, politics.[29]

In the meantime, a growing number of grassroots musicians, who did not belong to the underground scenes or the formal recording industry, also found ways to produce and distribute music recordings on the internet. Their work became known as "web songs" (网络歌曲) at the time. Typically simple pop tunes with low production quality, these songs circulated widely as MP3 files online and generated profits as mobile ringtones for a nationwide market. Their consumer base went far beyond young urban citizens who had been the primary listeners of Chinese independent music. "Web song" singers were seldom featured on "independent" self-releasing sites such as NeoCha and Douban. While a few stars emerged from this industry, the majority lacked access to both capital and professional marketing support, much like the more elite "independent" musicians.[30]

Since 2015, a new media ecosystem dominated by mainstream Chinese MSPs has brought further changes to the politics of self-releasing.[31] The two major MSPs, QQ Music and NetEase Cloud Music (NCM), inherited the Xiami model, integrating artist profile pages and self-releasing portals with interfaces for on-demand streaming. As self-releasing became platformized, the size and composition of the self-releasing musician group were further transformed. The newly formed platform-based system, established on a more (en)closed infrastructure, successfully incorporated both the group of "independent" musicians and the grassroots music workers in the contemporary version of the "web songs" industry, more recently revitalized by the rise of short-form video platforms; the boundaries between the two groups of music creators were also increasingly blurred. As a result, most of today's self-releasing musicians would not identify with "independent music" as a form of politics in opposition to the "mainstream." While many independent musicians still self-publish their work, the label "self-releasing artists" now extends far beyond those who would be considered "independent" in the past, although a more thorough analysis of the new politics of "self-releasing" is beyond the scope of this chapter.

Needs and Services: The Loss of Open Space for Artistic "Display"

In our focus groups, participants often espoused dual identities, seeing themselves both as producers of music and users of digital services. A commonly invoked term was "musician services," which encompasses a mix of basic services online, including an artist profile webpage with an independent domain name, streaming/download of uploaded music tracks, and connection with audiences and users through messages and comments. Following Western predecessors such as MySpace and SoundCloud, NeoCha, Douban and Xiami built localized self-releasing schemes based on their own existing services and infrastructures. Douban, for example, launched musician services in 2008 with artist profile pages, or "stations," that were conveniently linked to its vast, Discogs-like database of music records. For Xiami, self-releasing portals were integrated with its P2P-based MP3 download and streaming services, a design adopted by contemporary Chinese MSPs.

On the surface, aside from mainstream MSPs replacing websites and web portals as hosts for self-releasing services, the procedures for users who needed to publicly upload a track did not change much.[32] However, music creators active in the pre-platform era expressed a strong, nearly unanimous preference for the musician services provided by Douban and Xiami over those of QQ and NCM. Shanghai-based rhythm and blues singer C, for example, suggested that services provided by MSPs no longer suited her needs as a musician:

> In terms of my needs as a musician, streaming platforms today offer almost no help at all. Because I feel that in the early days, those sites, including Myspace, Sound-Cloud, Douban, and Xiami, their actual help and significance to me were much greater than NetEase and so on.

What, then, are the main services independent music creators need? This was a common topic of discussion in many focus groups. Particularly, the need to gain significant revenue from streaming was explicitly rejected by quite a few independent musicians, who stated that they understood self-releasing more as a process of "sharing" or "display" than as one of "selling" or "trade." For example, as C went on to say:

> As a musician, the biggest thing for me is that I need a platform for display, not just trading. . . . When I have made something, I hope my people can hear it there. It is not just for the purpose of selling. It is a space where I display my stuff as a musician.[33]

The idea of "display"—that is, showing in public a music creator's artistic identity through their recorded music works—was commonly regarded as an important need by independent music creators. As C's words imply, this conception often rests on a distinction between the creative value and commercial value of music, and ultimately between art and capital. From a classical Marxist perspective, this idea of "display" also points to a kind of musical labor that is not so alienated from creators. At a certain stage, digital self-releasing on early Chinese internet generated some potentialities, if not yet viable tools, to achieve this aspiration. Many participants agreed that website-based portals like Douban and Neo-Cha conveyed a sense of hope and emancipation, much like the discourse of "web utopianism" in the West.[34] Pertinent to independent music, these sites nourished a sense of community by enabling networks between individual music creators and their audiences, who shared similar aesthetics and, at times, political values. S, lead singer of a Chengdu postpunk group, recalled her early days self-releasing music on Douban:

> That experience was quite special. I didn't realize that I was listening to music made by my contemporaries. I didn't realize it at the time, but I feel that now more clearly. Now I cherish that feeling very much. Now, for example, NetEase Cloud or even Bandcamp cannot give me such a strong feeling.[35]

How exactly did early website-based self-releasing services afford such needs and aspirations? J, a rapper originally from Beijing, pointed out a bygone feature of the website-based Douban Musician system, which he dearly missed, called "open customization":

> Now that I think about it, I realize that Douban is a bit like a personal website, because the modules it contains can be customized. For example, there is a forum module, there is a streaming module, you can divide it into different albums, and then there are (modules for) pictures and videos. . . . There is a certain degree of open customization, but when I think about it, that is actually something from the Web 2.0 era, right? But think about its simplicity and its flexibility; it is true that none of today's music platforms can achieve that. I can't really do much on my NetEase Cloud profile page. Even

uploading works is quite complicated. . . . The room on these pages that can be customized and played with is very limited, and its visual presentation is also very poor. So now they are doing it in a very crude way, but why not try to do it better? There must be their own cost considerations and various things, which we have no way of knowing.[36]

J spoke as not only a musician services user but also an internet user. The quality of a website having room for customization and play results from what Zittrain terms "generativity"—that is, a system's openness toward "unfiltered contribution" or intervention—from bottom up by its users.[37] Zittrain sees generativity as a defining quality of the infrastructure of the open web, which allowed more space for the exercise of agency by its users as part of the system. In this sense, the loss of customizability in the shift from Douban to NetEase Cloud can be situated within broader trends of infrastructural (en)closure, as discussed above. Amid this historical shift, the relationship between listeners and the media through which they consume music was significantly reshaped; the active practices of "tinkering" afforded by the PC system could no longer be sustained in the "tethered" mobile devices, where users access media in a more fixed and standardized, though also more convenient, manner.[38] J's account illuminates a similar trajectory for self-releasing music creators, who lost the ability to design their own pages for the sake of creative, artistic display. Similarly, C expressed her frustration that contemporary MSPs no longer allow her to manage multiple side projects with one musician account:

> One of the mechanisms Douban established was that you can set up a small artist station, and this small station can have smaller substations. . . . With a general Douban Musician account, you could very conveniently manage these different substations, and then collect and manage data for different fan groups, and then display your different styles and different sides as an artist; this is a very, very musician-friendly service. . . . Now, for me, except for Douban, no platform can meet this particular need. Everyone here has different projects, different collaborations, and different directions of artistic expression, but today, if you want to achieve this, you might need to have ten mobile phone numbers to register for ten different accounts, and then you need to constantly log out and log in. . . . This is a very practical issue for me.[39]

The difficulty of managing multiple accounts as a self-releasing musician is further evidence that the platformized self-releasing system sacrifices openness for the efficiency of streamlined service. However, it also reveals a mismatch between independent music creators as users of self-releasing services and MSPs' imagined or anticipated users, as implied by their service designs. As many participants acknowledged, it is perhaps true that although the desire for "display"—showing an artist's diverse "sides" and aesthetic pursuits—is held dear by the independent music world, it may not be shared by all self-releasing musicians today. For example, music creators who produce "Douyin hit songs"—now working under increasingly anonymous conditions within "an extreme logic of flexible accumulation"—would likely prioritize attracting and monetizing "traffic" over such display.[40] Moreover, in the aftermath of the "copyright war," efficiently calculating and collecting streaming revenues became a top priority for Chinese MSPs. The platform-based musical system, with its streamlined DRM mechanisms, is far better equipped to cater to these needs than earlier web-based configurations.

As self-releasing services in China became increasingly dominated by MSPs, independent musicians found themselves no longer the primary group targeted by these services. As we argue, this resulted from the ability of mainstream MSPs to incorporate more forms of musical labor and bring music creators, previously outside the formal music industry, into their regimes of commodification. This shift echoes trends of music datafication discussed above, whereby the artistic value of musical commodities increasingly gives way to their value as datafied "content." In this sense, the changing value of recorded music on Chinese MSPs is embedded in the formalization of "informal circuits" within China's internet, which is inextricably linked to the expansion of informational capitalism in the country.[41]

Rights and Copyrights: (Re)Confronting Alienation

This process of formalization was also reflected in accounts of the changing routines and expectations of publishing demo tracks. H, an indie rock musician from Guangzhou, told us:

> Back then, when I uploaded music on platforms like Douban . . . I felt that I was sharing my music, and I had the right to control that music. . . . [I could] say, "This is a demo track; I think it is suitable to share it with everyone," so I upload it. It is really important for me that, one day in the future, when I have a more formal version of the track, and this demo is no longer suitable to be there, I can take it down. This seems to be a self-evident matter, but on today's music platforms like NetEase Cloud, it is not the case. You can't upload a song and delete it as you wish, so the thing [music] seems to have become something like [the platform's] product or asset. You have no choice.[42]

If the informal, website-based system smoothed the way for temporary display, or "sharing," of unpolished demo tracks, the challenge of removing these tracks represents another conflict between old and new self-releasing mechanisms. Whereas the old system facilitated artistic expression and social connections, the new system prioritizes the functioning of copyright law. In fact, although website-based self-releasing received almost unanimous praise, musicians in our focus groups admitted they had earned little to no revenue from Douban and only rarely from Xiami. In the early period of digitalization, self-releasing services were informal means of music distribution, often having little to no association with IP enforcement frameworks in China. After 2015, the few mainstream MSPs that survived the copyright war, having locked themselves into exclusive licensing deals with major music rights-holders, became solidly embedded in a new regime of internet governance with copyright protection at its core. This framework of governance is integral to the platforms' digital architecture and aggressively managed through DRM systems built into receiving devices. Within this framework, powerful institutional actors in the international and local recording industries—after a decade-long dispute with Chinese IT companies (most notably Baidu) over copyright infringement charges—allied with the now-dominant Chinese MSPs. As a result, formal copyright licensing has surpassed self-releasing as the prevailing means of digital music distribution in the age of streaming.

Today, on QQ Music or NetEase Cloud Music, the profile page of a superstar like Faye Wong—whose song rights are licensed to the MSP through publishing companies—does not appear so different from that of any independent music act who manages their own account through the platforms' self-releasing portal; in the platform-based ecosystem, they obey the same set of rules. In essence, self-releasing on contemporary MSPs is a process that enables the transfer of copyrights within defined technological and legal bounds. In this sense, any attempt to remove a previously uploaded track would be akin to attempting to terminate a signed record contract. In the platform era, agreements are automatically made during the registration and track submission process, as creators accept the platform's terms of use. Such "agreements" grant platforms the rights—often exclusive—to the recordings that creators upload. So H's view expressed above—that his musical works had become "products or assets" controlled by MSPs—is quite reasonable.

When tracks are submitted using the self-releasing portal on contemporary MSPs, the uploaded files are subject to an automated—or occasionally manual—copyright clearance check, a procedure that was uncommon in the pre-platform era. Though the screening criteria employed are never fully transparent, our participants identified "sensitivity," "originality," and "musicality" as among the most frequently cited justifications for rejection of their submitted tracks. Sensitivity—defined as whether a song contains any content deemed politically sensitive—reflects the further penetration of long-standing norms of cultural censorship in mainland China into the virtual space, in tandem with the process of copyrightization. Indeed, the state-led copyrightization campaign has been regarded as part of broader internet governance efforts, especially in the wake of the "Internet Plus" strategy launched in 2015.[43] One consequence of the more closed, formalized infrastructure of contemporary music circulation on MSPs is that music is increasingly subject to top-down censorship, often mediated through the platforms' own self-censorship. In our focus groups, many participants opined that censorship on MSPs was more common and stricter than ever before on the Chinese internet.

If "sensitivity" manifests state governance through the copyright regime, "originality" and "musicality" represent efforts by MSPs themselves to enforce copyright protection. Formal rights arrangements now form a fundamental basis of platform-based self-releasing systems in China. Participants recounted instances when they were asked to upload master recording files or screenshots to prove their authorship. Tracks identified as containing any copyrighted material—whether by means of song covers or sampling—were barred from the self-releasing system. However, our participants suspected that MSPs also acted out of a desire to block works submitted by "fake artists," which could negatively affect the distribution of streaming revenues, and that this lay behind the third screening criterion of "musicality." By imposing these criteria on the realm of digital self-releasing online—once praised for its democratization potential—MSPs are effectively using their power to define what counts as "music" through a new set of rules. These rules, which essentially stipulate what qualifies as an eligible and appropriate digital commodity under informational capitalism, drive a process of propertization of the once-informal musical work of self-releasing artists within contemporary China's platform economy.

These rules have, in turn, affected independent music-making. According to F, a Guangzhou electronic musician who records ambient and experimental music:

> [MSPs] will also examine the music itself. For example, if the music contains lengthy noise, or if it has very slight musical progression over time, then maybe, after they examine it through a who-knows-what kind of mechanism, it may not meet their requirements for submission. Especially for those of us who are doing ambient music, or music with fewer grooves or little musicality, it's quite a big obstacle.[44]

F's experience demonstrates how technological barriers, born from commercial imperatives and legal concerns, can have aesthetic consequences. Like the unfulfilled needs for customizable artistic displays discussed above, determination by platforms of what counts as music is another critical moment of confrontation in which independent music creators, previously accustomed to the loose guidelines of the self-releasing system in the early Chinese internet, encounter unexpected difficulties as they adapt to the rules of the new musical system. These changes, conditioned by a copyright-focused, platform-based ecosystem, are designed to meet the needs of the late-arriving but more powerful players who now control digital music publishing—namely, large rights-holders from the mainstream music industries. In other words, independent music creators have been forced to adapt to a new framework that serves the interests of "the majors," replacing an older system once used to circumvent them. The historical wheel has turned full circle.

A more dramatic confrontation between music creators and informational capitalist structures occurs when creators find their works uploaded to MSPs without their knowledge or consent. In the Chinese music streaming industry, it is no secret that MSPs employ web-scraping tools to "steal" content from one another's repertoire and other publicly accessible sources. This practice happened frequently during the copyright war era. Platforms openly adopt a passive problem-solving strategy, waiting for rights-owners to request the removal of songs rather than asking for permission beforehand. Such cases of "procedural (in)justice" abounded in our focus groups. Yet, when creators submit takedown requests to platforms, they are met with a profound moment of alienation. The first step for a takedown request requires that music creators set up a self-releasing musician account on the platform, and they must then prove that the music in question belongs to them. In our focus group, Shanghai techno musician M shared a disturbing vignette. When he tried to upload a song that he had already released on a foreign label to his NCM artist page, the platform rejected his submission. He was told that the track was already copyrighted. He later found that it had been uploaded to NCM under a different musician account, one automatically generated by the platform and most likely scraped from a foreign streaming site. As M lamented:

> This is equal to telling me that I was infringing my own copyright when I tried to upload music to my own account! This is equal to saying that I cannot prove that I am myself. Ridiculous—this is just so ridiculous.[45]

As an independent music creator, M was compelled to face a datafied, copyrighted version of himself—a "content producer" constructed by the platform-facilitated copyright regime. To reclaim rights to his own music, M had to prove that he was himself. That is, he had to conduct ethical work, subjecting his real self to the norms implied by the datafied self, agreeing to the terms and conditions dictated by the MSP, which required him to regard his work as IP available for trade. It is this kind of ethical work that habituates music creators to the new rules of the MSP self-releasing systems, pushing them into new spheres of datafied relations while simultaneously alienating them from their own works. Platformized self-releasing, in this sense, becomes a form of self-discipline, normalizing individual creators into actors in the structured and structuring field of informational capitalism. A few participants in our focus groups related such dilemmas to capitalist alienation, such as M:

> The platforms have opened a new era. The copyright issues and so on we mentioned earlier are basically those capitalist platforms treating you as a kind of asset, tricking and tricking you into putting your assets on their platforms as if they're their properties, through which they accumulate more wealth. . . . It is ultimately about how musicians escape the game of capital. Now, I can't think of any way to get out of this trap. . . . We have become a money-making tool for capitalists. We don't know when it happened, and we can't escape.[46]

For some independent musicians, however, the will to resist this form of alienation was the reason they identified with the politics of independent music in the first place. A veteran creator who has been active since the 1990s Chinese underground, M was fully aware of the irony here. Reflecting on his own past, M identified a regrettable change in mindset regarding musical independence:

I think the times do change our mentality. That is, the mindset of making music in the past was completely different from now, including your relationship with people in the music circle, your relationship with fans. . . . For example, before 2010, I was in the so-called independent or underground music circle. Everyone played together, and that was all I could think of. After 2010, suddenly, say, Douban came to me, saying, "How about I help you?" Back then, you had no idea that it was an intrusion of capital, and you had no idea what they wanted to do to you. . . . It was only in the past two years that we had figured out that, fuck, they are just trying to steal from our hands; they are just trying to grab our copyrights and use them to exchange for bigger money. This is what it boils down to.[47]

CONCLUSION

In this chapter, we have presented empirical observations from focus groups with Chinese independent musicians to interrogate and problematize self-releasing systems on contemporary MSPs. We have connected the present moment to the historical evolution of independent music distribution in China and, more broadly, to the historical path of music platformization. In the early 2000s, digital self-releasing was a means to evade the pervasive structures of the capitalist music industry. Music creators were empowered to work in a less alienated manner, with more freedom to adapt systems to their needs, and enjoyed greater control of their music at a time of loose IP regulation and open internet infrastructure. In the age of streaming, as digital self-releasing transformed from an emancipatory, connecting tool to a trap that constrains agency and artistic expression, independent music creators in China find themselves once again faced with the alienating logics of capitalism they sought to escape. These new, more restrictive self-releasing mechanisms are situated within the rise of closed platform infrastructures, datafied music distribution, and institutionalized copyright.

Due to space limitations, we have presented a somewhat linear historical narrative here. We have necessarily omitted some nuances and ambivalences that further complicate the story of independent music in China. Regardless, our account of the old and new dynamics of power and agency, with which Chinese independent music creators must cope, casts light on the entangled histories of music platformization in China and beyond. We hope to gesture toward more critical insights on music creators' perspectives and experiences in a new musical system under informational capitalism.

NOTES

1. Our findings rely on qualitative thematic analysis of focus group data. From November 2023 to February 2024, we recruited independent music creators in China to participate in seven online focus groups, consisting of five to six musicians, differentiated by genre. These focus groups discussed music

creators' perspectives toward digital platforms and their experiences of music production and distribution in the age of streaming.

2. Timothy D. Taylor, *Music and Capitalism: A History of the Present* (Chicago: University of Chicago Press, 2016).

3. Jason Toynbee, "Musicians," *Music and Copyright*, ed. Simon Frith and Lee Marshall (Edinburgh: University of Edinburgh Press, 2009), 123–38.

4. Peter Tschmuck, "Copyright, Contracts and Music Production," *Information, Communication and Society* 12, no. 2 (2009): 251–66.

5. David Hesmondhalgh, "Is Music Streaming Bad for Musicians? Problems of Evidence and Argument," *New Media and Society* 23, no. 12 (2021): 3593–3615.

6. Julie E. Cohen, *Between Truth and Power: The Legal Constructions of Informational Capitalism* (New York: Oxford University Press, 2019).

7. Keith Negus, "From Creator to Data: The Post-Record Music Industry and the Digital Conglomerates," *Media, Culture and Society* 41, no. 3 (2019): 367–84.

8. Robert Prey, "Performing Numbers: Musicians and Their Metrics," in *The Performance Complex*, ed. David Stark (Oxford: Oxford University Press, 2020), 241–58.

9. Nick Couldry, Aristea Fotopoulou, and Luke Dickens, "Real Social Analytics: A Contribution towards a Phenomenology of a Digital World," *British Journal of Sociology* 67, no. 1 (2016): 120.

10. José Van Dijck, Thomas Poell, and Martijn de Waal, *The Platform Society* (Oxford: Oxford University Press, 2018).

11. See for example the "Spotification" thesis in Jonas Andersson Schwarz, *Online File Sharing: Innovations in Media Consumption* (New York: Routledge, 2014); the "platform closure" argument in Robert Prey and Seonok Lee, "Platform Closure and Creator Creep: What We Can Learn from Korean Indie Musicians," *International Journal of Communication* 18 (2024): 4167–86; Julie E. Cohen, *Between Truth and Power: The Legal Constructions of Informational Capitalism* (New York: Oxford University Press, 2019); and David Hesmondhalgh, Raquel Campos Valverde, D. Bondy Valdovinos Kaye, and Zhongwei (Mabu) Li, "Digital Platforms and Infrastructure in the Realm of Culture," *Media and Communication* 11, no. 2 (2023): 296–306.

12. Jonathan Zittrain, *The Future of the Internet and How to Stop It* (New Haven, CT: Yale University Press, 2008); Tarleton Gillespie, *Wired Shut: Copyright and the Shape of Digital Culture* (Cambridge, MA: MIT Press, 2007).

13. Zhongwei (Mabu) Li and David Hesmondhalgh, "From P2P to the Cloud: Music, Platformization, and Infrastructural Change in China," *Chinese Journal of Communication* (2024): 1–17.

14. David Hesmondhalgh and Leslie M. Meier, "Popular Music, Independence, and the Concept of the Alternative in Contemporary Capitalism," in *Media Independence*, ed. James Bennett and Niki Strange (New York: Routledge, 2014), chap. 4; Negus, "From Creator to Data."

15. Negus, "From Creator to Data," 367.

16. Changwen Chen, Ziyi Wang, and Joel Gn, "Serve the Traffic: The Logic of Traffic and Its Politics on WeChat Platform," *Social Media and Society* 9, no. 3 (2023), https://doi.org/10.1177/20563051231193025.

17. Qian Zhang and Keith Negus, "From Cultural Intermediaries to Platform Adaptors: The Transformation of Music Planning and Artist Acquisition in the Chinese Music Industry," *New Media and Society* (2024), https://doi.org/10.1177/14614448241232086.

18. Tarleton Gillespie, *Custodians of the Internet: Platforms, Content Moderation, and the Hidden Decisions That Shape Social Media* (New Haven, CT: Yale University Press, 2018); Joanne E. Gray, *Google Rules: The History and Future of Copyright under the Influence of Google* (Oxford: Oxford University Press, 2020).

19. Zhen Troy Chen, *China's Music Industry Unplugged: Business Models, Copyright and Social Entrepreneurship in the Online Platform Economy* (Singapore: Palgrave Macmillan, 2021).

20. William Alford, *To Steal a Book Is an Elegant Offense: Intellectual Property Law in Chinese Civilization* (Stanford, CA: Stanford University Press, 1995).

source of income for some musicians in the twentieth century. However, the economic conditions of musicianship have been in transition since the beginning of the twenty-first century. With the advent and widespread use of digital technologies, especially the proliferation of high-speed broadband internet, which boosted the use of digital formats and rendered containers such as records, cassette tapes, and CDs largely redundant, the music industry had to confront the problem of unlicensed music (or piracy, if you will). The unlicensed and free sharing of copyrighted music via peer-to-peer (P2P) platforms was curtailed by the streaming model, but most musicians continue to struggle to earn a living from music, with recorded music providing only limited income.

The argument regarding whether musicians are in a better and a more sustainable financial position compared to the previous era of distribution technologies persists.[3] Some studies claim that "streaming has made earning a living from music more difficult" and that most musicians are not "able to make a living by [their] music alone, even though quite a few aspire for that."[4] To follow this aspiration, musicians either publish their music via TuneCore, CD Baby, and the like or make deals with labels working with the two biggest digital distributors in Turkey: The Orchard and Believe. In some cases, musicians choose to set up their own label and approach distributors for a deal. One of the most visible changes streaming has brought to the music industry in Turkey is that these distributors have replaced labels as bottlenecks. Distributors act as gatekeepers for most releases and any chance of promotion on streaming platforms. They also control the flow of revenue and data provided by streaming platforms. Distributors do not invest in musicians or offer much consultation other than how to use their own system, unlike pre-streaming labels, which at least provided studio time or sorted out some promotion. In this ecosystem, most "self-releasing" musicians in Turkey end up having to deal with most noncreative work themselves, while reinvesting what little they have earned back into equipment, studios, social media ads, and other expenses.[5]

On the consumption side, music streaming in Turkey is vibrant; it was the eighth-fastest-growing local market in the world in 2023.[6] However, this growth in volume is not reflected in overall revenues and musician earnings. A 2022 study estimated that streaming services pay an average of US$0.0004 per stream in Turkey.[7] This means if a single song is listened to one million times, the revenue transferred to rights owners is US$400. It is well-known that such rates are not very high globally (the highest estimate in 2022 belonging to Iceland, at US$0.0067), and they also fluctuate as they are calculated pro rata, depending on monthly and quarterly subscription and advertising revenues separately. For a musician to cover the monthly cost of living in Istanbul (US$617.30), they would need roughly 1.5 million streams. Meanwhile, a musician in Reykjavik only needs around two hundred thousand streams (to cover the average rent there of US$1307.70).[8] This presents further difficulties for those who choose to live in

Istanbul, the most expensive city in Turkey, where most of the music industry and the majority of music venues are located.

Due to these less-than-ideal streaming rates, whether in Iceland or Turkey, musicians are driven to depend on other forms of income, such as nonmusical work, live performances, merchandising, and, in some cases, other industries that utilize the work of musicians in commercial settings.[9] For musicians in Turkey, this can be even more difficult. While the current legal framework regulates copyrights and royalties, it lacks tangible regulations addressing the working conditions of musicians.[10] This creates an environment in which precarious and low-paid working arrangements are the norm for the majority. In a survey of 290 musicians and music industry workers conducted in Turkey in 2020, 62 percent of the respondents declared earning less than minimum wage, 63 percent had no social security, and 71 percent were looking for a "second job" in music to sustain their livings.[11] Relying on live performances, as most musicians do, has been problematic as well. As a part of COVID-19 precautions, live music after midnight was banned in Turkey on July 1, 2021. Seen by many as interfering with civil liberties and lifestyle choices, the ban lasted almost two years and was finally lifted on June 23, 2023.[12] This measure resulted in further financial obstacles for musicians and delayed recovery efforts from terrible conditions musicians faced during the pandemic, which, according to opposition member of parliament Gamze Taşçıer's statement on September 16, 2020, led to nearly one hundred musicians ending their lives due to financial troubles in the absence of live performance work.[13]

Recent conditions in Turkey, along with global and local effects of the COVID-19 pandemic, remain far from normal. Nevertheless, for most musicians, musicianship has never been a reliable form of self-employment.[14] Like most cultural workers, they have consistently faced conditions that have tended toward "insecure, contingent, and flexible work."[15] Musicians are often compelled to work as "solitary entrepreneurs" or "independent contractors," roles in which they receive little to no protection under conventional labor laws.[16] So poor conditions are hardly new or attributable to the streaming era—and, of course, many other workers face similar conditions in the "gig economy."[17] But in music, the contrast with the success of superstars, corporations, and businesses such as platforms, distributors, and big labels can be striking. What is arguably new in the streaming era is the intertwined economic aspects of musicianship and the social status associated with it.

Recent commentary has sought to address these and other changes. Simon Frith points to a long-standing refusal to view musicians as workers, a result of how artists perceive themselves in society and how society perceives them in return:

> The belief that music—making music—is in itself, fun, a pleasurable activity that shouldn't be thought of as work is embedded in our culture. Music is something humans do; we are all musicians—hence the vast number of amateur musicians, people who play for love. Such love of music is, of course, why people are willing to pay for musical labour in the first place, but it also means, perhaps, that they don't

really regard music as work. Its value is precisely as non-work. Musicians may, then,
be workers, but they shouldn't be![18]

The idea that musicians are workers who "shouldn't be" workers, along with the
notion of musicians as independent contractors or entrepreneurs, adds complexity to the economics of musicianship in the streaming era. But the specific case of
musicians who compose or perform for the advertising industry offers a fascinating example of worker-entrepreneur musicians. These musicians also raise questions of whether music is understood as craft or art, and whether performing and
composing commissioned music constitutes "selling out."

SELLING OUT TO THE AD INDUSTRY

David Arditi points to a common situation in which musicians tend to supplement their musical income through nonmusical or nonartistic jobs, with teaching
and freelance work serving as comparatively more artistic options.[19] Such work is
not usually considered selling out. But what if you pay the bills by making music
that requires "abandoning previously held political and aesthetic commitments for
financial gain"?[20]

Bethany Klein, Leslie M. Meier, and Devon Powers's account of selling out,
focused on concepts of musicians' autonomy and compromise, analyzes various
claims that selling out no longer exists.[21] They argue that the rise of skepticism
about the notion of selling out is based on the idea that "threats to established
revenue streams, especially record sales, have justified increasing involvement in
activities that previously would have been classified as selling out."[22] Klein also
points to changing cultural rationales, particularly the "omnipresence of marketing and branding for all communicative practice."[23]

Concurrently, the sociocultural status of advertising, particularly regarding
the relationship between arts and advertising, has been changing, and this is also
challenging the place of selling out discourse in the music industry.[24] Giana Eckhardt and Alan Bradshaw have claimed that views of musicians in advertising
have shifted—from being seen as selling out to being viewed as a "sought out" and
"nonproblematic" relationship benefiting all parties involved.[25]

In such conditions, where making music for ads is becoming socioculturally
less problematic and economically more viable than streaming revenues, understanding the perspectives and narratives of musicians who work for the advertising industry has the potential to provide valuable insight.

MAKING MUSIC FOR ADS

To begin, it is necessary to understand the difference between licensing previously published music and creating new, commissioned music for the ad industry.
Licensing or synchronization, as Eckhardt and Bradshaw claim, has become more

than a viable opportunity to reach more audiences and establish a career for "marginalized musicians."[26] This implies that, even though ad agencies or brands pay varying sums to license previously published music, exposure as payment is still a contributing practice. Meanwhile, Klein points out that for some ad campaigns, "product and creativity assuage concerns of selling out" for fortunate musicians, making it more than just about money and exposure.[27]

Working as a composer, producer, or session musician for commissioned jingles involves different circumstances. This work often includes a flat payment and offers very limited exposure that could lead to further paid work outside the ad industry. Admittedly, some singers in Turkey occasionally achieve fame via a frequently broadcast jingle, which they can then use to promote their own music via social media. The primary conditions for such work have not really changed. However, I argue that the significance of music commissioned for advertising has risen as an alternative source of employment for musicians by providing better conditions than the streaming economy. Since making jingles often pays better than other gigs, it has become a preferred source of income for many musicians.

As someone who worked as a musician in the ad industry between 2013 and 2015 in Istanbul, I found a job at a studio specializing in jingles and with strong connections to the ad industry to be financially beneficial. I was able to pay rent and bills for the first time by making music—just not my own music. The ad industry in the early 2010s seemed more lucrative and more vibrant than the music industry, which was in a period of transition following the expansion of Spotify into Turkey in 2013. By 2020 (the most recent available data), the online advertising market volume in Turkey was calculated by accounting firm PricewaterhouseCoopers to be US$600 million, while the broad category of "music, radio, and podcast" was only US$141.5 million.[28] Streaming revenues were only US$53.7 million by 2022.[29] Globally, the overall view is not very different, as the 2024 projections for total revenue in the music market is US$14.36 billion.[30] Meanwhile, worldwide ad spending is forecasted to surpass the US$1 trillion mark in 2024.[31]

These measurements of market volume or overall revenue tend not to reflect the full realities of actual people working in the industry, and they are often rather speculative. Furthermore, the data regarding the "ad industry" or "ad spending" are obviously not completely attributable to just music and musicians. Nevertheless, they suggest the greater prosperity of the ad industry. While the Turkish recording industries are recovering in the 2020s, it remains small compared to advertising, and streaming revenue cannot sustain most recording musicians.

AD MUSICIANS IN ISTANBUL

In the summer of 2023, I conducted interviews with ten musicians working for the ad industry in various capacities. The participants included a range of musicians, from session instrumentalists and singers to composers, producers, and studio owners, who actively create or perform music commissioned by ad agencies. These ten participants are insiders and key representatives with varying capacities of involvement in the business of making music for ads. As the interviews were conducted in Turkish, I translated the extracts presented in the following section, for which I made a diligent effort to ensure accuracy. I analyze their views below.

None of the participants took the view that streaming as an economic system was solely responsible for any major changes to their daily lives as jingle musicians. They discussed the lack of affordable recording equipment, the increase in the quantity of music (and the corresponding decrease in its "value"—with one participant likening streaming to an "open buffet" in a negative sense), and the arguably increased significance of promotion and social media as key drivers of change.

None felt it was possible to earn enough from streaming to pay their bills, at least in their current situation. When asked if it was possible for anyone to make a living from streaming, they often provided variants of remarks such as "Not unless you're Taylor Swift" or "Aleyna Tilki" (a Turkish pop star). One of the participants believed that success depended largely on the amount of music you have on Spotify, with five to ten albums serving as the threshold for earning real money, as well as on social media presence and promotion—both of which the participant, who is primarily a singer, chose not to partake in. One participant commented that musicians might survive on the basis of streaming if they had a strong sense of the zeitgeist and marketing, citing how the musician Jason Mraz's busking experiences taught him how to communicate effectively in a context of limited time and attention—skills that are also needed by jingle musicians. However, this musician also added that streaming can only work as a promotion for live gigs, even at the superstar level. This view was prevalent among all participants, who believed that concerts have become the main source of income for musicians who choose to bet on their own music, with streaming serving as an agent of exposure.

Another common view expressed by my interviewees was that if they could earn the same amount of money from their own music, they would not be creating jingles. Attempting a shift back to making their own music was seen either as a risk not worth taking or a dream that needed to be supported with other income. Understandably, all of them sought opportunities to make money from advertising due to the difficulty of making it in the music industry in the first place.

Finding work in the ad industry as a musician is tied to a network of a handful of big media music production companies in Istanbul. These are run by early adopters who started making music for ads in the 1990s. Around this core group, which can charge ad agencies and advertisers enough to afford session musicians, are small studios and bedroom producers who often use MIDI instruments and only hire singers—who, at least for now, cannot be easily replaced by virtual studio technologies. All six of the composers/producers I interviewed had varying degrees of connections to the ad industry when they started, some having worked in ads in other positions and others having close contacts in the industry—friends, family, and the like. Session musicians, however, seem to follow a different path, entering the industry by either actively reaching out to large ad music production companies or being headhunted by them. In both cases, the strength of their networks seems to heavily influence their chances of entering the industry, as well as the amount of work they find once they are in.

For my interviewees, the most prominent differences between making music for the ad industry and doing so for the music industry seem to revolve around the questions "Was it commissioned?" or "Whose music is it?" When composing, performing, or producing commissioned music for the ad industry, participants felt they were involved in a craft rather than an artistic endeavor. Nevertheless, echoing Timothy Dean Taylor's claim that "selling out is no longer an issue," all participants believed that making music for ads is essentially no more "selling out" than other forms of music-making.[32] One said that "Mozart also did commissions"; another spoke of how orchestral musicians don't compose their own music; another spoke of how even if you self-publish your own music, "Spotify is your boss." The consensus was that everyone is trying to make a living and that being paid for music always involves an element of selling out. Nevertheless, some participants pointed out that musicians who primarily work in the recording industry, even if they are not making their own music, tend to see jingle musicians as "second class," "fabricator musicians who should not receive royalty," or even "not musicians." One composer/producer participant, who previously tried to make it with his band before starting to create music for ads, had a hard time recruiting their musician friends for jingle gigs because those musicians "see it as a service, like you're lowering your ego a step."

All but one participant considered themselves to be "workers." The only musician who rejected this notion was a session musician who stated, "Those who see themselves as workers stay that way, but those who see themselves as musicians always go up." The same participant, while thinking of themself as an entrepreneur, did not believe most other musicians are entrepreneurs, since they lack understanding, awareness, and action regarding their rights as musicians. Two more participants were hesitant about seeing musicians as entrepreneurs: one associated entrepreneurship with innovation, which they expressed is not necessarily involved in making music for ads, and the other attributed such a title only to musicians with a business mindset, who are often involved in alternative ways of earning money from music, such as ads.

One of the main determinants of musical work in advertising, similar to most other creative work in Istanbul, is turnaround time. For most projects, the production pace can be much faster than in the recording industry. It also involves many intermediaries demanding the time of the musicians: the production company, ad agency, and marketing communication department of the advertiser. The workflow timeline is set by the advertiser, with the broadcast date and time being mostly unmovable deadlines, especially if it is a TV commercial. This hierarchical structure and the looming deadline mean that musicians often have only a couple of days—or sometimes even less—to write, compose, perform/record, and produce. While this creates a suboptimal condition for quality music production, once the musicians are accustomed to the pace, they can come to prefer it, as it means more money for less time spent.

This was one of the most prominent patterns in the views of the participants when they were asked about the advantages of making music for the ad industry. Since most ad work requires a considerably short piece of music—often less than a minute for TV commercials—once it is composed, it can be recorded in a relatively short time. For a session musician or singer, the initial recording session usually takes only a couple of hours—sometimes even less—and, if the jingle does not receive any revisions from the production company, ad agency, or client, that can be it. Nevertheless, this does not necessarily mean that it is not a labor-intensive process. As one of the participants—a producer and a media music pioneer—put it, once the newcomers attracted to the "easy money" enter the industry, they see that "that [a] thirty-second piece of music isn't made in thirty seconds." Furthermore, to streamline the process and deliver within the given time, many people are involved or, for comparatively smaller studios, one or two people perform multiple roles, playing instruments, recording themselves, managing client relations, maintaining the studio, and paying taxes.

For all ten participants, making music for ads was the primary—or at least one of the primary—sources of income, and the common view was that it is also financially preferable to other ways of making money from music. Furthermore, considering that the ad industry in Turkey, as mentioned above, is more vibrant in comparison and that project turnarounds are much shorter than in popular music production, ad gigs are seen as more achievable.

However, to enter the industry, one has to become a part of the existing network and abide by the traditional power relations in which it operates. Uploading a song to Spotify does not require such social networks, and in this respect, it can be said that participation in ad music-making is less "democratic." Nevertheless, the participants interviewed seemed to prefer expending social effort to become part of the ad music network rather than the labor of standing out among a myriad of others on streaming platforms. This preference went beyond "better than nothing." For some, ad work was "simply better," and one participant went as far as to state that "advertising provides a very good economic platform for creative workers."

One other key aspect of being a creative worker in the ad industry is the requirement to let go of one's own aesthetic preferences, at least enough to stay relevant in the industry. As a musician, it is not feasible to be known for your competence in just one genre—something producers and session musicians may also face in the music industry, albeit to a lesser extent. All my interviewees spoke of the multi-genre, multi-culture, and multi-instrumentalist requirements of their work. One framed this as an advantage, stating that "you really have to learn music" to survive as a jingle musician. Another saw it as a welcome challenge. Additionally, the three singers I interviewed spoke of the challenges of singing performance in ad work, such as the need to sometimes provide a "romantic vibe," a "parental" feeling, or a cheerful or manic feeling. One of these participants, with a degree in musical theater and acting, expressed that the work was "fun and exciting."

Making music the way other people want is central to music advertising work in Istanbul. One of the participants said that this is "the clearest drawback" of their profession since they are "writing made-to-order music—you don't get to say something like, 'I woke up feeling like this, so I'll write something like that.'" Another participant stated that they never felt "so glad [they] composed that" about a jingle. For session musicians and singers, it seems that an even smaller area of artistic expression is allowed. As they work with composers, producers, and arrangers, the freedom of expression diminishes further, and they end up "doing whatever they want," "not leaving the confines of the brief," and having "no say in" how they perform. For both composers and performers, just being flexible in terms of genre and style is not enough; they also have to be flexible about their involvement in artistic decisions, since, as one participant noted, "you'll have a difficult time, if you're strict about the music you make."

Apart from artistic aspects of the conditions of being an ad musician, concrete practical difficulties also need to be faced. Musicians must always be available. While this is a larger issue and also applicable to most of the ad industry in Istanbul, it frames how ad musicians experience their working hours and work-leisure balance. Since most projects involve fast turnarounds, creative workers in the industry are always on call, ready and waiting. This is common across the industry. Some larger studios can negotiate more reasonable timelines, holidays, or working hours, but smaller studios and freelance musicians often struggle to reject the unrealistic demands of ad agencies or clients. One participant warned me before we started the interview that "if the phone rings," they would have to do a revision. Another complained that they cannot plan any holidays since they do not know when they will be free. Two of my interviewees told me that they go on vacation with a bag of recording equipment, and one had to build a make-shift recording booth in their hotel room while their friends were enjoying the beach. On the other hand, two other participants noted that some projects allow them to plan their working hours, and they enjoy being able to decide when they work and when they don't.

While the industry demands constant availability and timely delivery for most of its workers, the payments they receive are not always punctual. Apart from one session musician participant, who gets paid the same day due to their insistence on such a practice, all my interviewees mentioned the issue of late and sporadic payments. Although there are informal conventions regarding payment terms since most freelance musicians and producers do not sign contracts with agencies or clients that include binding clauses for payment times, most musicians find themselves pursuing their payments, sometimes for up to three months. Considering that Turkey has been undergoing an economic crisis with considerably high inflation rates for many years, late payments constitute a serious problem, as the sum shrinks with each day.

One participant characterized their overall economic situation as "the unbearable weight of being a musician in Turkey." Admittedly, some of these issues are not specific to musicians, let alone musicians working for the ad industry. One such issue is political and economic unpredictability and instability. An established studio owner told me that they "managed to keep the shop open for all these years, in a country like Turkey where you don't know what'll happen tomorrow—I can't even predict whether we'll be able to keep it up." This situation in Turkey adds to the increasingly "insecure, contingent and flexible work" conditions of musicians in the streaming era, and even a financially preferred occupation for a musician shows significant signs of precarity.[33]

On the other hand, one upside, according to most participants, was the continuity and volume of the production of advertisements. The boom in Turkish media means that "media is constantly being produced . . . and someone needs to compose music for all that . . . and this means sustainability. I think it's more secure in that sense"—that is, more secure than work in the streaming-dominated music industry. In a manner embodied in this participant's quote, musicians who work in the ad industry may feel better about betting on the future of the advertising and media industries, where music is utilized, than on the contemporary music industry in the streaming era. Though, naturally, they all expressed hopes and suggestions to better their conditions.

POSSIBLE IMPROVEMENTS

When asked about what can be done to make their profession less precarious, one of the most prominent and common answers was standardization of contracts, turnaround times, working hours, and, most importantly, fees. Participants were vocal about the need for such standardization, particularly for fees, both to protect themselves from ruthless negotiations and to prevent price-cutting, something newcomers feel they need to do. Musicians also sought the enforcement of existing intellectual property and copyright laws, which are poorly understood by musicians in the industry. Finally, and possibly most significantly, the need to "do it together," "organize," and "unionize" was expressed.

A newly formed organization called the Media Music Composers Association (known in Turkey as Medya Müziği Bestecileri Derneği, MMBD) represents an intention to provide collective solutions to some of the problems experienced by participants. One notable aspect of MMBD is the distinct use of the term "media music composer." While the name implies that it concentrates more on the composers and producers, with limited emphasis on representing performers and session musicians, the association defines itself as a legal entity that represents composers, arrangers, producers, and musicians who create music for film, TV, ads, games, and so forth. It seeks to provide sectoral definitions that will end conceptual confusion in the industry, legal support for members, standardized licensing agreements and letters of consent, and access to shared information, ideas, and resources.

However, in its current state, it is arguable whether MMBD has much collective power, beyond the individual power of influential members, to instigate such standardization and pave the way for better conditions for media musicians. For one interviewee, "at the end of the day, it's not a union, and since it's not a union, it's only a declaration of will." However, "it reminds us that we're not alone in the sector. I think this is a really nice feeling."

One of the most practical functions of MMBD in its current state is the support experienced members can provide for newcomers. One participant and MMBD member stated that they aim to "plot a route for newcomers so they don't feel like a fish out of water," since, as another participant concurs, "the newcomers tend to have a lot of questions," because they know how to make music but not how to handle the noncreative work required by the profession. The association and the communication network provide "knowledge and experience transfer" and, according to one participant who did not necessarily have that when they started, "it's priceless."

CONCLUSION

In today's streaming-dominated music industry, the socioeconomic and practical conditions of musicianship are more congruent with musicians as craftworkers than as artists. That is due to the economic conditions laid out by the platform economics of the streaming era, which seem to create its hegemony over cultural industries; selling music as a functional product, rather than selling music as art, becomes more socially visible in the case of media musicians in Istanbul. In such circumstances, what needs more attention is this transformation, as well as the conditions of the musicians who are undergoing it to become media musicians. Klein's analysis of the convergence between advertising and music regarding "the reliance of artists on corporate patrons" and how this gives way to "greater corporate influence," resulting in a representative lack of "space for all sounds and messages," is a compelling answer to the question of how to approach this phenomenon.[34]

Nevertheless, the socioeconomic conditions of the streaming era and the transformation of the nature of musicianship force artists to navigate this increasingly commercialized landscape, often without having the opportunity to reflect critically on the underpinnings of their practices.

Contextualizing their experiences and viewpoints within the broader economic and cultural environment contributes to a deeper understanding of the intricate dynamics between streaming revenues, musicianship, and the advertising industry. While streaming has created the challenges of earning a livelihood as a musician, the economic conditions that fostered streaming have also facilitated a shift in the industry, pushing many musicians to seek alternative methods for financial sustainability. In such conditions, the pull of the advertising industry as a means of making steady money seems to become more visible, especially when the increased significance of advertising is considered. For the musicians I spoke to, notions of "selling out" were largely dismissed, and the weight of economic realities prevailed, particularly in the face of Turkey's ongoing political and economic uncertainties. Nevertheless, tensions surrounding the perceived status of ad musicians persist, pointing to deeper conflicts regarding labor and creativity in the streaming era.

NOTES

1. Chris Nickell, "Promises and Pitfalls: The Two-Faced Nature of Streaming and Social Media Platforms for Beirut-Based Independent Musicians," *Popular Communication* 18, no. 1 (2019): 1–17; Ruth Towse, "Dealing with Digital: The Economic Organisation of Streamed Music," *Media, Culture and Society* 42, no. 7–8 (2020): 1462.

2. David Kusek and Gerd Leonhard, *The Future of Music* (Boston: Berklee Press, 2005); Stephen Witt, *How Music Got Free* (New York: Penguin, 2015).

3. David Hesmondhalgh, "Is Music Streaming Bad for Musicians? Problems of Evidence and Argument," *New Media and Society* 23, no. 12 (2020): 1–23; Anja Nylund Hagen, "Datafication, Literacy, and Democratization in the Music Industry," *Popular Music and Society* 45, no. 2 (2021): 1–18.

4. Towse, "Dealing with Digital," 1462; Saskia Mühlbach and Payal Arora, "Behind the Music: How Labor Changed for Musicians through the Subscription Economy," *First Monday* 25, no. 4 (2020), https://doi.org/10.5210/fm.v25i4.10382.

5. Taylan Ozan Ceyhan, "The Working Conditions of Self-Releasing Musicians in Turkey" (master's diss., Middle East Technical University, 2023), 87.

6. Luminate, *2023 Year-End Music Report* (Los Angeles: Luminate, 2024), https://luminatedata .com/wp-content/uploads/2024/01/Luminate_Year-End_Report_2023.pdf.

7. "How Much Do I Get per Stream on Spotify (2022 Edition)," *iGroove*, May 24, 2022, https://www .igroovemusic.com/blog/how-much-do-i-get-per-stream-on-spotify-2022-edition.html.

8. "Cost of Living in Istanbul," Numbeo, n.d., accessed April 12, 2024, https://www.numbeo.com /cost-of-living/in/Istanbul?displayCurrency=USD; "Cost of Living in Reykjavik," Numbeo, n.d., accessed April 12, 2024, https://www.numbeo.com/cost-of-living/in/Reykjavik?displayCurrency=USD.

9. David Tilson, Carsten Sørensen, and Kalle Lyytinen, "Platform Complexity: Lessons from the Music Industry," *IEEE Xplore*, January 1, 2013, https://doi.org/10.1109/HICSS.2013.449.

10. Ceyhan, "Working Conditions," 110.

11. Selda Dudu, Evrim Hikmet Öğüt, and Özge Ç. Denizci, *Türkiye'de Müzik Emeğinin Durumu: Türkiye'deki Müzik Emekçilerinin Çalışma Koşulları ve Gelir Durumları Üzerine Araştırma Raporu* (The state of music labor in Turkey: A research report on the conditions and income status of music workers in Turkey) (Ankara: Avrupa Birliği Sivil Düşün Programı [European Union civic thought programme], 2022).

12. Haber Merkezi [Newsroom], "Müzik Yasağı Türkiye Genelinde Kaldırıldı" (Music ban lifted nationwide), *Medyascope*, June 23, 2023, https://medyascope.tv/2023/06/23/muzik-yasagi-turkiye -genelinde-kaldirildi/.

13. Kültür Servisi, "CHP Milletvekili Taşçıer: Pandemi Süresince 100'E Yakın Müzisyen Intihar Etti" (RPP MP Taşçıer: Close one hundred musicians committed suicide during the pandemic), *Evrensel*, September 16, 2020, https://www.evrensel.net/haber/414255/chp-milletvekili-tascier-pandemi-suresince-100e-yakin-muzisyen-intihar-etti.

14. Jan Drahokoupil and Brian Fabo, "ETUI Policy Brief N° 5/2016 European Economic, Employment and Social Policy: The Platform Economy and the Disruption of the Employment Relationship," European Trade Union Institute policy brief, July 14, 2022, https://www.etui.org/publications/policy-briefs/european-economic-employment-and-social-policy/the-platform-economy-and-the-disruption-of-the-employment-relationship.

15. Rosalind Gill and Andy Pratt, "In the Social Factory? Immaterial Labour, Precariousness and Cultural Work," *Theory, Culture and Society* 25, no. 7–8 (2008): 3.

16. Jonas Andersson Schwarz, *Online File Sharing: Innovations in Media Consumption* (New York: Routledge, 2018); David M. Arditi, "Precarious Labor in COVID Times: The Case of Musicians," *Fast Capitalism* 18, no. 1 (2021): 13.

17. S. Craig Watkins, *Don't Knock the Hustle: Young Creatives, Tech Ingenuity, and the Making of a New Innovation Economy* (Boston: Beacon Press, 2019).

18. Simon Frith, "Are Workers Musicians?," *Popular Music* 36, no. 1 (2016): 115.

19. Arditi, "Precarious Labor," 18.

20. David Hesmondhalgh, "Indie: The Institutional Politics and Aesthetics of a Popular Music Genre," *Cultural Studies* 13, no. 1 (1999): 34.

21. Bethany Klein, Leslie M. Meier, and Devon Powers, "Selling Out: Musicians, Autonomy, and Compromise in the Digital Age," *Popular Music and Society* 40, no. 2 (2016): 222–38.

22. Klein, Meier, and Powers, "Selling Out," 225.

23. Bethany Klein, *Selling Out Culture, Commerce and Popular Music* (New York: Bloomsbury, 2020), 124.

24. Bethany Klein, *As Heard on TV: Popular Music in Advertising* (Abingdon: Routledge, 2016).

25. Giana M. Eckhardt and Alan Bradshaw, "The Erasure of Antagonisms between Popular Music and Advertising," *Marketing Theory* 14, no. 2 (2014): 167–83.

26. Eckhardt and Bradshaw, "Erasure of Antagonisms," 168.

27. Klein, *Selling Out Culture*, 120.

28. "Global Entertainment and Media Outlook 2023–2027," PricewaterhouseCoopers, n.d., accessed April 1, 2024, https://www.pwc.com.tr/global-entertainment-media-outlook-2023-2027.

29. "Türkiye Müzik Endüstrisi" (Turkish music industry), Bağlantılı-Hak Sahibi Fonogram Yapımcıları Meslek Birliği (Turkish phonographic industry society), 2021, https://tr.mu-yap.org/wp-content/uploads/2022/05/TURKIYE-MUZIK-ENDUSTRISI-2021-2-pdf.io_-1.pdf.

30. "Music—Worldwide | Statista Market Forecast," Statista, n.d., accessed April 1, 2024, https://www.statista.com/outlook/amo/app/music/worldwide.

31. Statista, "Advertising—Worldwide | Statista Market Forecast," Statista, n.d., accessed April 1, 2024, https://www.statista.com/outlook/amo/advertising/worldwide#:~:text=Ad%20spending%20in%20othe%20Advertising.

32. Timothy Dean Taylor, *The Sounds of Capitalism: Advertising, Music, and the Conquest of Culture* (Chicago: University of Chicago Press, 2014), 229.

33. Gill and Pratt, "In the Social Factory?," 3.

34. Klein, *Selling Out Culture*, 124.

12

Inequity by Design

Music Streaming Taxonomies as Ruinous Infrastructure

Raquel Campos Valverde

Despite the prevalence of music streaming for the past fifteen years, we still know very little about the design and architecture of its software infrastructures and their impact on society. Cultural studies scholarship highlights how music streaming platforms (MSPs) can reproduce geographical disadvantages,[1] as well as the differences between overtagged content—often white, male, Western artists—and undertagged genres,[2] which inevitably leads to higher exposure and revenue for some. However, the role of specific technologies, and the political-economic forces shaping them, has been insufficiently explored. In previous work, I and others have highlighted the contradictions of existing research on algorithmic music recommendation.[3] Despite critical algorithm studies of classification in music streaming and algorithmic recommendation,[4] it is still unclear which taxonomies of music are used by MSPs. Previous research highlights that algorithmic tools are heavily influenced by the corporate culture of each platform and the individual understandings of those who work in these companies.[5] Employees' influence in categorizing and marketing music is also confirmed by studies that address the role of human editorial curation.[6] Moreover, streaming platforms have not adopted established notions of diversity or the common good used by public service media.[7] The top-down model that has come to represent music streaming calls for further investigation of the ways these taxonomies are produced, distributed, and infrastructurally crystallized.

SOFTWARE AND INFRASTRUCTURES: A FRAMEWORK
FOR STREAMING INDUSTRY RESEARCH

Approaching software and coding technologies as pieces of music streaming infrastructure builds on previous research calling for attention to systems of internet architecture, to understand how media content is organized and distributed.[8] I conceptualize taxonomical systems as infrastructure because they are critical elements to the existence and technical functioning of recommender systems and streaming products such as playlists. Considering digital infrastructures beyond "the [physical] stuff you can kick,"[9] in Lisa Parks's words, I contribute to software and critical data studies.[10] Paraphrasing Parks, I consider music streaming software infrastructure to be the stuff you can click—or the stuff (specific commands and seed queries) you are expected to type when you want to stream music—meaning the necessary digital building blocks on which routes for music exploration are designed. In paying attention to the "logical infrastructure"[11] of MSPs, I demystify the standards and protocols used for music distribution, marketing, and consumption. Also approaching streaming taxonomies from an anthropological notion of digital infrastructure, I concur with Nick Seaver's understanding of algorithms as sociotechnical structures.[12] I also further problematize the park ranger metaphor uncovered in his ethnography among software developers,[13] whereby the designers of MSPs see themselves as tour guides in the wilderness of infinite musical choices. With this metaphor, developers present themselves in a kind light as friendly rangers who escort clueless picnic-goers. It obscures, however, developers' role in creating the maps, routes, trails, and enclosures that users and audiences follow, absolving them of responsibility for what park plots people visit—and therefore which areas produce revenue—as well as how nature is presented to visitors.

Understanding taxonomical software infrastructure as a navigation map or route through the musical wilderness, here I question the existing digital architecture of music streaming to critically evaluate what is available to the public, how it is designed, for whom, and how politico-economic decisions are made. In doing so, I follow the postcolonial cultural economy approach of Anamik Saha,[14] combining a cultural economy approach to media circulation with a postcolonial approach to race and culture. Therefore, this chapter combines an investigation of the political economy of streaming with an analysis of its cultural politics. I investigate what musical taxonomies are currently used by streaming services, along with the kinds of cultural visions and understandings of music cultures inscribed in these taxonomies. With this, I shed light on how musical taxonomies in streaming services might influence music cultures—for instance, by demonstrating that music streaming taxonomies contain encoded Western biases as engineered forms of cultural imperialism.

To examine these issues, I collected empirical material from three sources. First, interface analysis of six music streaming platforms (Spotify, SoundCloud,

Apple Music, Tidal, YouTube Music, and Amazon Music) and critical analysis of the metadata coding standards used to create musical taxonomies in the industry. Second, discourse analysis of PR materials produced and distributed by the aforementioned streaming platforms, as well as industry talk at seven music industry conferences: Music Biz (United States), Music Ally Next (United Kingdom), by: Larm (Norway), Reeperbahn and CTM (Germany), and MIDEM (France). Third, interviews with industry stakeholders and, to a lesser extent, researchers and members of public institutions and nonprofit organizations.

INEQUITY IN THE MUSIC INDUSTRIES

The research took place simultaneously with an increase in racial conflict visibility that heavily influenced industry discourse. Between 2020 and 2021, inequity and specifically anti-Black racism was at the forefront of much industry discourse and PR. As the *Music Industry Action Report Card* of the Black Music Coalition[15] points out, many organizations pledged a number of equity initiatives,[16] particularly addressing poor workforce diversity.[17] However, many of these initiatives only produced modest results. Ambitious corporate statements did not match practical action. For instance, YouTube claimed, "We now have an umbrella of work that seeks to address racial justice, equity and inclusion while *embedding that into the fabric of how we operate. We want to ensure that we're dismantling structures and not creating systems that just reproduce bias.*"[18] However, the company's initiatives focused on racial inclusivity among content creators[19] and did not address the technical fabric of recommendation or catalog. Other companies, such as Apple and Warner Music Group, did not present specific equity strategies or programs. Inequity was also largely absent from the agendas of the biggest music conferences in 2023; only Music Biz had a specific conference track about it, with multiple sessions. Similarly, Music Biz was the only event that had a specific conference track about metadata reform, but this was treated as a revenue and rights management issue distinct from inequity. There are considerable differences between industry PR discourse and the companies' internal strategies, and industry discourse on racism and equity initiatives lacks any cultural understanding that would lead to a systematic reform of musical taxonomies and streaming metadata. The technical aspects of music distribution are largely absent from these discussions, assuming that access and representation can solve existing racial injustice. As Saha points out, the media industries often focus on getting representation "right."[20] Going beyond industry–co-opted considerations of representation, my interest here is to reanimate scholarly interest in questions of cultural imperialism and global flows of music distribution in the streaming era. In line with Saha and Mel Stanfill,[21] I focus on the productive power of digital media interfaces in reinforcing specific social logics. The first section of this chapter defines taxonomies and the types of taxonomies I address.

The second section explains the relationship between taxonomies and metadata coding standards in the industry, and how these reproduce inequity. The third section provides further evidence of these practices in curatorial decisions. The last section provides a critical interpretation of these findings.

WHAT'S IN A TAXONOMY?

To understand how content is currently organized by music streaming services, first the notion of taxonomy should be further contextualized. By musical taxonomies, I mean the classification systems of music genre, mood, instruments, and other musical elements that provide the software infrastructure for navigation, product design, distribution, and recommendation in streaming platforms. Each recommender system is thus based on a multiplicity of musical taxonomies. Indeed, the concept of taxonomy cannot be understood as a static, singular one (i.e., *a* taxonomy), but rather as an assemblage of *taxonomies*. Music is also a paradigmatic case of competing taxonomical classifications based on cultural and power differences,[22] so the notion of taxonomy carries strong historical and ideological connotations. Originally used in the natural sciences, it has an aura of neutrality, implying that taxonomies are just organized representations of reality. Yet, no classification system can be devoid of ideological value. The concept is also closely related to the classification of the natural world by Western powers during colonial expansion.[23] Comparative musicology and ethnomusicology have often emphasized the taxonomical indexing of music and the search for universality in music,[24] as seen in works like the Cantometrics project[25] and the Hornbostel-Sachs classification for organology.[26] The current use of the concept in music research derives from its applications in computer science.[27] However, computer science studies fail to question the mood/activity/genre triad favored by streaming platforms or the Western understandings of those categories. More culturally aware research to measure and classify music based on predefined essential attributes comes from music information retrieval (MIR). However, taxonomical efforts based on Western music descriptors lack the necessary nuance to provide fair recommender systems for all, and much research is still required before these are redesigned with a more international scope.[28]

More critical perspectives on taxonomy and recommender systems come from media and communication studies. Amelia Besseny[29] stresses the importance of folksonomies, understood as users' own classifications of genre and content, as well as the unequal relationship between taxonomy and folksonomy in streaming interfaces, where curated expertise provided by the platforms has gradually become more prominent as these recommender systems became central to monetization strategies over time.[30] However, user-generated data may lead to problematic categorizations that reproduce power structures or homogenize diverse music genres.[31] Such data might also reflect users' reproduction of certain keywords for their own commercial benefit.[32] However, reinforcing the apparent duality of taxonomy versus folksonomy as equivalent to expert versus amateur forms of knowledge would not be helpful. On the contrary, in the next sections I show how the music streaming taxonomies currently used in metadata transactions—presented as scientific efforts of encoding—are often based on folksonomies as well, namely the preexisting folk understandings of music within the platforms themselves, created by workers and stakeholders in the music industry.

Popular music studies often discuss taxonomies by analyzing genre. Since the 1990s, scholarship has dealt with the commercialization and distribution of popular music in an increasingly global market, highlighting the politics of classifying and indexing music from around the world.[33] These studies foreground issues arising from a Western-centric vision of musical discovery and "audio tourism," particularly prevalent in contested catchall categories such as "world music," which draws an artificial line between the musical "West" and the "Others." Seaver has revived this debate by arguing that the classifications used for diverse cultural sounds in recommender systems stem from a Western vision of the designer and listener as placed in the global center of musical knowledge.[34] However, this Western-centric sociotechnical practice within streaming industries is by no means a new development. The growth of the recording industry at the beginning of the twentieth century included the creation and consolidation of categories such as "foreign music" and "race music," conflating cultural stereotypes with the listening communities imagined and marketed.[35] Similar developments are apparent in the streaming era.[36]

Academic research and streaming PR suggest a decline in the centrality of genre. Streaming platforms now focus on categories such as mood and genreless playlists to market their music catalogs.[37] However, genre continues to be a crucial element in music streaming success, with hyperpersonalization and curation based on genre in the past few years.[38] As Seaver points out, designers of recommender systems pigeonhole listeners into categories, only to later try to relativize them.[39] In September 2023, Spotify introduced "Daylist," a hyperpersonalized playlist product that provides recommendations to users based on niche mood and microgenre combinations,[40] drawing on the work of data scientist Glenn McDonald, who was laid off shortly after the product's launch.[41] The logic behind such products is not to abandon genre completely, but to understand genre as a dynamic element, seemingly organized in a nonhierarchical way for recommendation, as well as increasing product differentiation aligned to the platform's marketing campaigns. However, this seems a rather unachievable target for algorithmic coding. The understanding of genre in the centers of power where streaming is designed and developed may be loaded with questionable ideas of otherness and discovery. Problems with streaming diversity and inequity in music streaming stem from platforms' limited consideration of genre definitions and the critical deployment of taxonomies, whether as a theoretical or a technological concept.

TAXONOMIES AND METADATA

The process of adding catalog to an MSP can be divided in three stages, where taxonomies are applied to the metadata of digital music files. First, metadata creation and collection; second, standardization; and third, editorialization and curation (fig. 12.1). Digital file metadata describe the content of the file and its relationship to other content, in ways that algorithmic systems read for recommendation and discovery. In the current model, a potentially data-rich but disparate, nonstandardized catalog from an artist, record label, distributor, or aggregator is tagged and organized, either manually or via an automated service provider at the creation stage. At the second stage the catalog is made metadata-compliant, and its data simplified following coding standards overseen by the Digital Data Exchange, a nonprofit, industry-funded organization. Lastly, music files are sent to MSPs for curation and editorialization, where new metadata tags are applied to catalog, effectively destandardizing and re-branding it, but sometimes reinstating its initial data richness. In the case of user-generated content, whether from a self-releasing artist or a fan, metadata may remain exactly as rich or poor as in the original upload, and platforms do not always intervene to standardize or optimize it.

Here I argue that the metadata coding standards currently followed by the music industry fall short of any systematic understanding of genre classification or sound analysis, particularly outside the Western canon. Huge disparities exist between how different genres can be categorized and therefore how they underpin editorialization, marketization, and eventually monetization in the recording industry. For example, the current genre list of the Digital Data Exchange[42] (reproduced in Box 1)—the most widely used set of coding standards for over ten different kinds of metadata in the music industry applied at the second stage of standardization before the catalog is added to MSPs[43]—has a list of fifty-eight subgenres for the parent label "Rock." However, it only includes nineteen subgenres for a vast area such as "Latin" (specifically focused on Mexico), and a meager five for "Reggae."[44] This reflects the history of this genre list as a piece of Western-centric classification, developed from a previously existing list used by a major record label. This seed list did not include much information about relationships between genres, unlike a treelike library of genres and subgenres or a network-style taxonomy.

The DDEX coding standard also allows for greater granularity when it comes to Western genres. For example, it distinguishes "Classical Music" as a Western genre, "Classical" as a subperiod of Western art music, and classical in terms of structure. However, it only anticipates coding "Hindustani Classical Music" as a subgenre of "Indian Music" or "Pakistani Music." The genre definitions in the data library that works as a manual for distributors and aggregators also stem from the

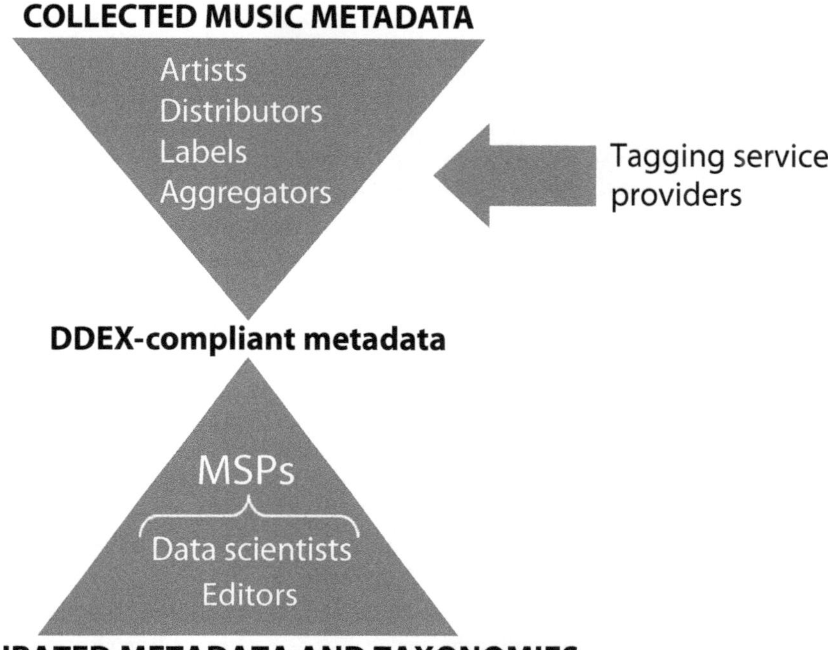

FIGURE 12.1. Three stages of applying musical taxonomies to the metadata of digital music files. Generated by the author.

major label–provided seed list, not following any specific scholarly sources or dictionaries. Asked about these asymmetries (specifically on classifying cha-cha-cha music as "Traditional" and not "Latin"), a DDEX representative responded:

> DDEX has to rely on its members to provide the information, and as usual, if you have a certain slant . . . so [record label] is a label which makes most of its money with traditional pop. Well, then, of course, that will be their focus. And the Latin bit. Yes, [record label] has a big Latin set of labels. But that's less the dance music, more the, the rhythm-and-blues kind of Latin music, I would think. So that's where their focus is. That means that the . . . especially the classical music, especially the ballroom dance kind of music that doesn't make a lot of money, therefore it gets less, um . . . attention. Therefore, it will be underdeveloped. Not good, but there you go.[45]

This overreliance on commercial interests contrasts with the treatment of genre or geographical areas in musicological sources such as *The Grove Dictionary of Music and Musicians*.[46] Moreover, this careless codification of genre in the DDEX standard is even more striking when compared with relatively underresourced efforts, such as the taxonomical maps circulated in fan communities.[47]

BOX 1. DDEX GENRE CODE LIST

Blues
A Genre characterized by a loose narrative lyrical style, use of call-and-response, the blues scale and blue notes, a small set of common chord progressions, and trance-like walking basslines. Originated in African-American communities in the Deep South of the United States in the late 19th century.

ClassicalMusic
Traditional Western art music. Though wide-ranging in sound and style, it is largely characterized by its system of staff notation, and often by its musical complexity.

CountryMusic
A Genre characterized by the use of Guitar and twangy vocals. Instrumentation traditionally includes any of drums, bass, Banjo, Fiddle, Harmonica, ElectricOrgan, or steel guitar, though much modern music makes heavier use of Pop and Rock instrumentation. Originated in the Southern United States in the 1920s and influenced by southern Folk music tradition, including Blues and descendant styles of Scottish, Irish, and English folk traditions.

ElectronicMusic
Music created primarily by electronic Instruments and methods, including manipulation of both digital and circuitry-based forms of audio technology.

Folk
A term that refers both to the traditional folk music of the British Isles and of North America (typically the music of the people, as opposed to ClassicalMusic—the music of royal courts, aristocracy, and the well-to-do) and to modern genres which primarily take influence from those traditions (particularly during and after the 20th century folk music revival).

Gospel
Sung Christian music with roots in traditional Hymns and early African-American spirituals. Often features call and response, and often performed a cappella, with FootStomps and HandClaps for rhythmic accompaniment. Gospel can also feature Piano, Organ, Guitar, drums, and other Instruments.

HipHop
A Genre that typically features rapped vocals (emphasis on rhythm over melody, characteristically verbose compared to other musical styles) over beats. It emerged out of neighborhood block parties as part of a broader hip-hop culture among African-American communities in the Bronx in New York City in the late 1970s.

Jazz
Wide-ranging Genre characterized by the use of swung rhythms, blue notes, polyrhythms, and particularly, extensive improvisation. It incorporates a wide range of influences, from Blues, Ragtime, and ClassicalMusic (particularly that of Impressionist composers such as Debussy), to spirituals and West African cultural and musical traditions. It first emerged as the

	Dixieland style of music among the African-American communities of New Orleans in the late 19th and early 20th centuries. Throughout the 20th century, it developed stylistically across the entire United States, from Kansas City to New York City.
Latin	An umbrella Genre that encompasses most music from Spanish or Portuguese speaking areas of the world.
Pop	Popular music, for lack of a better term. Consists almost entirely of short-to-medium length songs, with heavy use of verse-chorus structures and a strong emphasis on melodicism and catchiness. Has no singular sound—often incorporates the popular sounds of the day (thus pop was synonymous with Rock through the 60s, picked up elements of EDM in the late 2000s/early 2010s, and often features trap beats in the late 2010s).
R'n'B	Originally a marketing term for popular African-American music with a strong beat, R&B has since come to define a few specific styles that are perhaps as much sonic as racial categories. The term has several distinct associated sounds, depending on the era. In the early 50s, R&B described popular Blues, records, and in the mid-50s, the term came to denote Gospel and Soul music, as well as popular styles with elements of electric blues, acoustically similar to contemporary Rock-NRoll (which itself grew out of early R&B). In the 70s, it largely referred to Soul and Funk, and in the 80s, the term began to refer to a sonic hybrid of earlier R&B, Pop, Soul, Funk, rap, and ElectronicMusic. It has morphed and evolved while maintaining this hybrid identity to the present day, taking on newer production and performance styles as time passes.
Reggae	A Genre that features an offbeat staccato feel, halftime one drop drum grooves, and socially conscious lyrics. Influenced by mid-century American RAndB and Jazz, Jamaican Ska, and traditional Jamaican music such as mento. Emerged in Jamaica, particularly around Kingston, in the late 1960s.
Rock	Song-focused, typically ElectricGuitar-centric and beat-driven Genre that emerged in the 1940s and 50s from Blues, RAndB and CountryMusic. Many variants and styles exist, though most feature at least ElectricGuitar, bass, drums, and a lead singer.
Spoken	Primarily non-musical and focused on the spoken word.
Traditional	Folk and court music traditions outside of North America and the British Isles.
UserDefined	A Type of an Entity which is defined by a sender of a Ddex-Message in a manner acceptable to its recipient.
WorldMusic	A fusion of various western popular music Genres with different global Folk music styles.

Other details of the DDEX standards are also worth mentioning. For example, the current coding standard uses English script and is not optimized for the inclusion of special characters such as an umlaut or a tilde (ä, ö, ü, ñ, etc.), let alone the use of languages other than English. Although multiple alternative names can be input for an artist depending on the region, this creates disparities in the way files are traced and therefore monetized. In practice, the only way to resolve these genre and language problems is manually inputting user-generated metadata.

A further example of this Western-centric approach to music tagging comes from the taxonomy list of Musiio,[48] an MIR and AI-based automated metadata tagging provider for important industry players such as SoundCloud (now its parent owner) and Sony, often used at the first metadata stage. Musiio's eighty-four-item taxonomy map (reproduced in Box 2) is simplified to such an extent that it hardly provides any granular data for complex musical territories such as "Indian" (at the time of writing, a single genre tag trained on Bollywood music),[49] and it only identifies fourteen different instruments, all of them based on Western musical terminology.

An employee of a tagging service provider explained that technical developments typically respond to the needs of their industry clients while trying to remain independent of any specific MSP (rather than copying their taxonomies).[50] For example, as most clients are expanding into India and Latin America, tagging services are likely to increase granularity for those targets. Confirming Jeremy Wade Morris's analysis of Pandora,[51] what is considered "exotic" or secondary in technical development depends on a Western- and English-centric perspective. While automated systems such as Musiio (SoundCloud) and Echo Nest (Spotify) may analyze inherent elements of a song such as rhythm, decisions about what to add, when, and how much these factors should carry in business decisions are ultimately human choices. In other words, metadata and taxonomy development follow the business culture of the tech industry, whereby underdeveloped products are launched in the consumer market for live testing, and then progressively modified following market trends. In comparison, public institutions like the British Library use more rigorous metadata standards, employing an established genre list adopted from the US Library of Congress.[52] However, public institutions focus more on digitizing historical recordings than cataloging new musical trends, and thus their taxonomies may not always be transferable to streaming services. Another contrasting example from a private business comes from the Nigerian start-up Josplay, a contextual and editorial metadata company that provides metadata services for application developers.[53] Josplay is currently developing an open-source African Music Library that aims to target this gap in the market with more granular metadata maps for the African continent. However, placing private businesses in charge of generating this metadata does not seem to be a sustainable, long-term solution.

BOX 2. MUSIIO TAXONOMY MAP

GENRE

Afrobeat	Folk	Death Metal
Afropop	Funk	Mandopop
Classic Blues	Gospel	J Pop
Blues Rock	Old School Hip Hop	Pop Rock
Classical Instruments	Alternative Hip Hop	80s Pop
Classical Vocals	Trap	Electro Pop
Classic Country	Pop Rap	Contemporary Pop
Bluegrass	UK Grime	Punk Rock
Country Pop & Rock	Indian	Ska Punk
Disco	Indie Rock	Smooth R&B
Adult Contemporary	Indietronica	Alternative R&B
Smooth & Vocal Jazz	Industrial	Dance R&B
Downtempo	Swing	Reggae
Ambient	Bebop	Dub
Synthwave	Jazz Fusion	Dancehall
House	Salsa	Rock & Roll
EDM	Reggaeton	Classic Rock
Techno	Latin Pop	Hard Rock
Trance	Classic Metal	J Rock
Breakbeat	Heavy Metal	Alternative Rock
Drum & Bass	Thrash Metal	Early Soul
Dubstep	Nu Metal	Neo Soul
Hardcore	Metalcore	

INSTRUMENT

Banjo	Guitar	Synth
Bass	Keys	Trumpet
Beats	Percussion	Violin
Brass	Piano	Woodwind
Drums	Strings	

Similar trends exist in other domains of taxonomical management, such as quality control. Taxonomical review and updating at DDEX are managed by its members and stakeholders—144 institutions, including record labels and publishers, unequally distributed around the world, with only a few participating in the technical management group. The DDEX membership fees and the cost of sending staff to meetings excludes many independent record labels and distributors. As of this writing, only four members are based in Latin America. This also implies that quality control is not managed by a team of music experts as such, besides those recommended by the industry. Relatively small teams also work in companies such as Musiio and Josplay. However, Josplay employs some musicologists and area experts to contribute to quality assurance. Even the British Library, which has in-house curators for each region covered by its catalog, is severely understaffed in this respect. At the 2023 Metadata Summit, DDEX and music industry representatives acknowledged inaccuracies in 5–10 percent of the catalog, mostly in content from indies and self-releasing artists, whom the industry aspires to train or eradicate from the data value chain. However, this seems to be an underestimation, since the British Library admits inaccuracies in about a third of its catalog, particularly in content from major commercial players.[54] Moreover, if the coding standards are optimized for English-language and Western music genres, the system leaves minor industry players to do the heavy lifting in terms of quality control. Even if the argument for metadata reform is considered solely in terms of monetization, the current system appears to be unfit for purpose. Previous metadata reform trials by Universal and Amazon Music have demonstrated that richer metadata increases music usage.[55] In short, the industry sees software development as stemming from a center located in Europe or North America and gradually extending to the rest of the world. In light of this encoded taxonomical inequity, I argue that, in the current digital music industry landscape, music metadata can be considered unequal by design, as the value and importance assigned to certain information is preestablished from the initial data input by the developers of the technology and the relative power of the institutions involved in the process.

TAXONOMIES AND CURATION

These inequity issues persist in the third stage of the taxonomical indexing process, during the editorial phase. The already Western-centric metadata received by the streaming platforms are further modified and adapted by data scientists working to standardize the content within a specific platform. Moreover, the editorial team may not have input into, or much knowledge of, the preceding process. A senior editor at a major streaming service admitted to not knowing what the DDEX coding standard was, and did not think it relevant.[56] They gave more importance to the internal "deep metadata" creation provided for them by their tagging provider than the metadata received from artists or distributors when a new track is pitched to curators. If anything, they saw their role as reconciling these two sets of metadata and making sure songs were editorialized correctly to maximize revenue.[57] Therefore, at any point in the distribution process, the same track file can be classified under at least two distinct taxonomical systems, which here I name metadata taxonomies (during the first stage) and editorial taxonomies (during the third stage). Moreover, the genre and mood mapping of specific employees, such as Glenn McDonald, may have more influence over the taxo-nomical systems of streaming platforms than the industry-sanctioned standards or the artists themselves.

MSPs' PR materials highlight the human character of the curatorial and recom-mendation process to convey expertise and product differentiation.[58] However, a curation-dependent taxonomical system has further implications. A curator with-out specific area expertise may tend to group many non-Western musics under the "world music" or "pop" categories or prioritize the organization and display of non-Western musics that reproduce "clickbait orientalism."[59] For example, Spotify proposes playlists such as "Spanish Tapas Bar," consisting of a mix of flamenco and fusion in multiple languages (including a track titled "Gypsy Flame") or "Tulum Vibes," with an equally mixed bag of genres, languages, and titles like "Salsa Cali-ente." Asked about curatorial influence on platforms, an informant at Josplay shared that MSPs simply lack expertise for relatively simple editorial tasks, such as writing PR copy that distinguishes Afrobeat as a genre from Afrobeats as a family of genres within a wider cultural network.

Problems with representation existed in the music industry long before the advent of streaming, traditionally dominated by certain powerful groups in terms of class, gender, and nationality. But even if diversity hires and the equity initia-tives discussed above made a significant impact, this still leaves aside the lack of accountability in this new gatekeeping system. A considerable number of tasks in these processes remain divided among separate teams and divisions, or are com-pletely subcontracted, with little holistic vision of editorialized products. Curato-rial trends also affect representation, currently oscillating between genreless or genre-fluid curation and hyperniche genres.[60] These editorial practices introduce a significant level of destandardization, but this is ignored in industry conversations about metadata reform. A streaming service informant highlighted the difficulty of reconciling these trends with good metadata practices:

> The openness of the audience, especially younger users raised on streaming, they don't see the importance of a lot of the genre labels. . . . So some of these kinds of arbitrary orders are coming down, which I think is a really good thing. . . . Challeng-ing from a metadata perspective, though . . . and I think in a lot of ways the metadata systems have always been very niche. . . . Gracenote, they have the top level, and then each layer down becomes like a root system of ever-expanding subgenres. I think across the industry, we're leaning more into those subgenres. I don't know if that's good necessarily.[61]

There are also significant contradictions in approaching user-generated content as a problem to solve by training DIY artists, while simultaneously absorbing and monetizing user playlists, despite their inclusion of problematic folksonomies. In the Spotify case highlighted by the Anti-Defamation League,[62] a fashwave playlist was simply absorbed from a user account, effectively monetizing fascist content and prompting the question of who the target audience is. Asked about such extremist instances, a streaming informant admitted:

> It's dangerous. It's really not good. . . . There's a lot of those that [are] floating around and . . . there's new ones created all the time. There's really not a huge amount of oversight. . . . It's just trusting these systems and saying that people will understand that it's just the system creating this. But I don't think that's the case. . . . There should be a team that's reviewing this, a QA team. . . . It's just the scale is so huge. . . . They shouldn't be . . . promoting this music. I think it's one thing for them to exist. I think it's another thing for them to be actively serving them to users.[63]

However, I do not wish to reinforce the distinction between expert versus user-generated content. As Morris highlights, such a reductive dichotomy ignores the fact that algorithms and recommendation engines are fundamentally human in their construction and execution, and as this section shows, a combination of human and machine-coded factors is at the root of this engineered inequity. Despite the difficulty of observing these slants and omissions in action, examining the interface provides rich clues about the practical consequences of these flawed sociotechnical systems and human decisions behind them. For example, Spotify, which seems to have a penchant for inappropriate copy, suggests a playlist for "Sahara" under the tagline "The hottest music from the hottest desert" with a cover image from desert blues band Tinariwen, whose song "Toumast Tincha" appears in the list. Yet the tagline seems wholly misguided for a political song protesting the Mali government. Similarly, Tidal's "Geography for Beginners" section includes country-specific playlists for Scotland, Wales, and Ireland, but not one for England, which seems to be taken as the default position of the listener. In the same way that music marketing in the past was designed for a white listener in the West to explore "world music,"[64] these lists are created with a white, English-speaking audience in mind, encouraging exploration of other, more or less exotic content—whereas a specific Western geography is not subject to this kind of exploration.

In light of this evidence, the overall picture of taxonomical indexing and recommender systems is not positive. The current system lacks a systematic understanding of taxonomy, let alone genre or organology, and those involved in it often lack the expertise to manage it, with little understanding of the roles of other stakeholders. A great deal of the decision-making to date has been improvised as the technology developed, with responsibility placed on machinic entities such as algorithms, or, at best, the workers at both ends of the process, such as artists and curators. My contention here is that the combined effects of (a) the lack of a systematic metadata infrastructure, (b) the lack of diversity in the music industry workforce in general, and streaming in particular, and (c) the use of folksonomies and social data produce forms of inequity that are encoded in streaming recommendations from the start. Streaming taxonomies "platform racism"[65] in music, creating, distributing, reproducing, and amplifying existing social inequities.

STREAMING TAXONOMIES AS RUINOUS INFRASTRUCTURES

Spotify's first TV advertisement in the United States in 2013 specifically correlated the platform's existence with positive social impact. However, it did so from a Western-centric perspective. An intense young masculine voice poetically stated:

> Why can a song change the world? Because music is a force for good, for change, for whatever. . . . It lives inside us, because we were all conceived to a 4/4 beat.[66]

Two years later, Tidal's launch focused on discourse around social justice and fair revenues for artists. The campaign foregrounded Black US artists with the mottos "Tidal for all" and "Tidal puts the power back into the artist's hands." Artist and co-owner Alicia Keys spoke of it as "a moment that will forever change the course of music history."[67]

In *The Promise of Infrastructures*, Hanna Appel, Nikhil Anand, and Akhil Gupta[68] approach technological infrastructure from an anthropological perspective, where human discourse and material structures are intertwined in the production of cultural objects. They posit that

> the material and political lives of infrastructure frequently undermine narratives of technological or social progress, drawing attention instead to the shifting terrain of modernity, distribution, inclusion, and exclusion in most of the world. . . . New infrastructures are promises made in the present about our future. Insofar as they are so often incomplete—of materials not yet fully moving to deliver their potential— they appear as ruins of a promise.[69]

In this chapter, I have followed the same approach and analyzed streaming taxonomies as the result of discourse about streaming classification and curation and the software infrastructures available to encode those projected musical values. In doing so, I argue for understanding music streaming taxonomies as ruinous infrastructures. Here, ruination does not mean decadence but, rather, a state of in-betweenness between what it promises and what is delivered. The promises made on behalf of these infrastructures evoke normative notions of common good, access, inclusion, and equality. Yet these infrastructures are not designed to deliver these potentials. MSPs developed as an enclosure of the generative possibilities of earlier principles of internet architecture, offering a both-and solution for music consumption in the digital era.[70] This model would satisfy rightsholders and record labels while simultaneously enhancing audiences' experiences of music online by providing access to vast catalogs of music. Indeed, the word "'promise' implies that a technological system is the aftereffect of expectation; it cannot be theorized or understood outside of the political orders that predate it and bring it into existence."[71] I have evaluated the taxonomical realities delivered by platforms to compare them with the promises of recommendation and discovery. In doing so, I show that musical taxonomies and recommender products are digital infrastructures that "show the making and management of difference—class, race, gender, religion, and beyond—in the technics and politics of everyday life."[72] That is, streaming infrastructures such as taxonomies are forms of governance of the politics of music, but also forms of politics in themselves.

This focus on streaming infrastructures such as taxonomies is crucial for understanding the formation of audiences and publics, because "publics can be gathered or forestalled by the materials of infrastructure"[73] themselves. If, in the words of Jeremy Wade Morris, "the legitimacy of infomediaries, in the rhetoric of those who create and employ them, is based both on the cultural knowledge of those creating the databases and algorithms, but also on the size and scope of the databases and the efficacy of the algorithms themselves,"[74] at this stage of development, the role of musical taxonomies and MSPs as intermediaries of music experience has to be firmly questioned. This is not to return to tired arguments about human versus machine production of culture. Ultimately, all products of machines are also products of human design and intervention, and music curation has always been produced through the interaction between humans and some kind of music technology. However, from an audience studies perspective, it is crucial to pay attention to the structural elements shaping music consumption, in ways that assign political responsibility to the humans that design digital infrastructures. Instead of further encouraging the implementation of beta products for general consumption, as is common in computer science, a critical social science perspective should lead the development and design of technological infrastructures. Taxonomical development is a kind of generative technology in the sense that it creates the categories and classifications for recommendation and discovery of music, and as such, it is culture- and world-making. Streaming taxonomies, despite their partial invisibility, are thus sociotechnical political orders structured around particular ideas of race and culture. These taxonomies are implemented on and absorbed by society and should be approached with cultural and politically informed perspectives. Instead of considering recommender systems as an expert-only framework and decrying the consequences of their design, we should question why they are designed this way in the first place.

NOTES

1. Tamas Tofalvy and Júlia Koltai, "'Splendid Isolation': The Reproduction of Music Industry Inequalities in Spotify's Recommendation System," *New Media and Society* 25, no. 7 (2023): 1580–1604.

2. Tom Johnson, "Chance the Rapper, Spotify, and Musical Categorization in the 2010s," *American Music* 38, no. 2 (2020): 176–96.

3. David Hesmondhalgh, Raquel Campos Valverde, D. Bondy Valdovinos Kaye, and Zhongwei (Mabu) Li, "The Impact of Algorithmically Driven Recommendation Systems on Music Consumption and Production: A Literature Review," UK Centre for Data Ethics and Innovation Reports, 2023, https://ssrn.com/abstract=4365916.

4. Arnt Maasø and Hendrik Storstein Spilker, "The Streaming Paradox: Untangling the Hybrid Gatekeeping Mechanisms of Music Streaming," *Popular Music and Society* 45, no. 3 (2022): 300–316; Nick Seaver, *Computing Taste: Algorithms and the Makers of Music Recommendation* (Chicago: University of Chicago Press, 2022); K. E. Goldschmitt and Nick Seaver, "Shaping the Stream: Techniques and Troubles of Algorithmic Recommendation," in *The Cambridge Companion to Music in Digital Culture*, ed. Nicholas Cook, Monique M. Ingalls, and David Trippett (Cambridge: Cambridge University Press, 2019), 63–81.

5. Seaver, *Computing Taste.*

6. Tiziano Bonini and Alessandro Gandini, "'First Week Is Editorial, Second Week Is Algorithmic': Platform Gatekeepers and the Platformization of Music Curation," *Social Media and Society* 5, no. 4 (2019), https://doi.org/10.1177/2056305119880006; Tiziano Bonini and Paolo Magaudda, *La musica nell'era digitale* (Bologna: Il Mulino, 2023).

7. Andres Ferraro, Gustavo Ferreira, Fernando Diaz, and Georgina Born, "Measuring Commonality in Recommendation of Cultural Content: Recommender Systems to Enhance Cultural Citizenship," in *Proceedings of the 16th ACM Conference on Recommender Systems (RecSys '22)*, ed. Jennifer Golbeck et al. (New York: Association for Computing Machinery, 2022), 567–72.

8. David Hesmondhalgh, Raquel Campos Valverde, D. Bondy Valdovinos Kaye, and Zhongwei (Mabu) Li, "Digital Platforms and Infrastructure in the Realm of Culture," *Media and Communication* 11, no. 2 (2023): 296–306.

9. Lisa Parks, "'Stuff You Can Kick': Toward a Theory of Media Infrastructures," in *Between Humanities and the Digital*, ed. Patrik Svensson and David Theo Goldberg (Cambridge, MA: MIT Press, 2015), 355–73.

10. Lev Manovich, "Data," in *Critical Terms in Futures Studies*, ed. Heike Paul (Cham, Switzerland: Springer, 2019), 61–66.

11. Brett M. Frischmann, *Infrastructure: The Social Value of Shared Resources* (New York: Oxford University Press, 2012).

12. Seaver, *Computing Taste*, 7.

13. Seaver, *Computing Taste*, 141.

14. Anamik Saha, *Race, Culture and Media* (London: Sage, 2021).

15. Naima Cochrane, *Music Industry Action Report Card, 2022* (Nashville: Black Music Action Coalition, 2023), www.bmacoalition.org/musicindustryreports.

16. TuneCore, *Be the Change: Gender Equality in Music* (Brooklyn: TuneCore, 2024), www.datocms-assets.com/119568/1715870701-eng-be-the-change-report_2024.pdf; "The Keychange Pledge," Keychange, accessed March 22, 2024, www.keychange.eu/the-keychange-pledge; "Equal Access," mtheory, accessed March 22, 2024, http://mtheory.com/.

17. "Reports," Pub Royalty Queen, accessed April 16, 2025, www.pubroyaltyqueen.com/reports.

18. Cochrane, *Music Industry Action Report Card*, 23, my emphasis.

19. Cochrane, *Music Industry Action Report Card*, 23.

20. Saha, *Race, Culture and Media.*

21. Saha, *Race, Culture and Media*, 276; Mel Stanfill, "The Interface as Discourse: The Production of Norms through Web Design," *New Media and Society* 17, no. 7 (2015): 1059–74.

22. Patrick Savage, "Alan Lomax's Cantometrics Project: A Comprehensive Review," *Music and Science* 1 (2018), https://doi.org/10.1177/2059204318786084.

23. Patricia Fara, *Sex, Botany and Empire* (Thriplow: Icon Books, 2004); Jonathan Marks, "The Long Shadow of Linnaeus's Human Taxonomy," *Nature* 447, no. 28 (2007), www.nature.com/articles/447028a.

24. Patrick Savage, Steven Brown, Emi Sakai, and Thomas E. Currie, "Statistical Universals Reveal the Structures and Functions of Human Music," *Proceedings of the National Academy of Sciences* 112, no. 29 (2015): 8987–92.

25. Alan Lomax, *Cantometrics: An Approach to the Anthropology of Music* (Berkeley: University of California Extension Media Center, 1976).

26. Eliot Bates, "The Social Life of Musical Instruments," *Ethnomusicology* 56, no. 3 (2012): 363–95.

27. Ashton Anderson, Lucas Maystre, Rishabh Mehrotra, Ian Anderson, and Mounia Lalmas, "Algorithmic Effects on the Diversity of Consumption on Spotify," in *WWW '20: Proceedings of The Web Conference 2020*, ed. Yennun Huang, Irwin King, Tie-Yan Liu, and Maarten van Steen (New York: Association for Computing Machinery, 2020), 2155–65; Bruce Ferwerda, Emily Yang, Markus Schedl, and Marko Tkalcic, "Personality and Taxonomy Preferences, and the Influence of Category Choice on the User Experience for Music Streaming Services," *Multimedia Tools and Applications* 78 (2019): 20157–90.

28. Babat Nikzat and Rafael Caro Repetto, "KDC: An Open Corpus for Computational Research of Dastgāhi Music," in *Proceedings of the 23rd International Society for Music Information Retrieval Conference, ISMIR 2022, Bengaluru, India, December 4–8, 2022*, ed. Preeti Rao et al. (Geneva: Zenodo, 2022), 321–28; Ajay Srinivasamurthy, Sankalp Gulati, Rafael Caro Repetto, Xavier Serra, "Getting Started on Computational Musicology and Music Information Research: An Indian Art Music Perspective," in *Indian Art Music: A Computational Perspective*, ed. Preeti Rao, Hema A. Murthy, and S. R. M. Prasanna (New Delhi: Scheme for Promotion of Academic and Research Collaboration, 2023), 3–38; Thomas Nuttall, Miguel G. Casado, Andres Ferraro, Darrell Conklin, and Rafael Caro Repetto, "A Computational Exploration of Melodic Patterns in Arab-Andalusian Music," *Journal of Mathematics and Music* 15, no. 2 (2021): 168–80.

29. Amelia Besseny, "Lost in Spotify: Folksonomy and Wayfinding Functions in Spotify's Interface and Companion Apps," *Popular Communication* 18, no. 1 (2020): 1–17.

30. JohnVPandora, "Pandora Curator Profiles Are Live!," Pandora Community, September 20, 2023, https://community.pandora.com/t5/Creators-Blog/PANDORA-CURATOR-PROFILES-ARE -LIVE/ba-p/135514.

31. Eric Leonel Cureño, "Negotiating Artistic Representation in the Era of #Worldmusic: Trends, Challenges, Authenticity, and the Artist's Perspective" (master's diss., Arizona State University, 2021), www.proquest.com/openview/6f9df3895548b90740cd8c315a57c7f8/.

32. Sophie Bishop, "Influencer Management Tools: Algorithmic Cultures, Brand Safety, and Bias," *Social Media and Society* 7, no. 1 (2021), https://doi.org/10.1177/20563051211003066.

33. Georgina Born and David Hesmondhalgh, eds., *Western Music and Its Others: Difference, Representation, and Appropriation in Music* (Berkeley: University of California Press, 2000); Veit Erlmann, "The Politics and Aesthetics of Transnational Musics," *World of Music* 35, no. 2 (1993): 3–15; Steven Feld, "Pygmy Pop: A Genealogy of Schizophonic Mimesis," *Yearbook for Traditional Music* 28 (1996): 1; Anahid Kassabian, "Would You Like Some World Music with Your Latte? Starbucks, Putumayo, and Distributed Tourism," *Twentieth-Century Music* 1, no. 2 (2004): 209–23; Mark Slobin, *Subcultural Sounds: Micromusics of the West* (Hanover, NH: University Press of New England, 1993); Timothy Taylor, *Global Pop: World Music, World Markets* (New York: Routledge, 1997).

34. Seaver, *Computing Taste*, 83–85.

35. David Brackett, *Categorizing Sound: Genre and Twentieth-Century Popular Music* (Oakland: University of California Press, 2016).

36. Amara Pope, "How Canadian R & B Artists like Drake and Justin Bieber Complicate Ideas of Race, Music and Nationality," *The Conversation*, October 12, 2023, http://theconversation.com/how -canadian-randb-artists-like-drake-and-justin-bieber-complicate-ideas-of-race-music-and-nationality -212240; Johnson, "Chance the Rapper."

37. Mads Krogh, "'Context Is the New Genre': Abstraktion Og Singularisering i Digital Musikformidling," *Norsk Medietidsskrift* 27, no. 3 (2020): 1–15; Elias Leight, "Future 25: John Stein, Co-Creator of Spotify's Genre-less Pollen Playlist," *Rolling Stone*, September 30, 2019, www.rollingstone.com/pro /features/future-25-john-stein-co-creator-of-spotifys-genre-less-pollen-playlist-889139/.

38. Tatiana Cirisano, "Pandora's Mobile App Gets a Hyper-Personalized Revamp," *Billboard*, October 1, 2019, www.billboard.com/pro/pandora-refreshes-mobile-app-personalized-for-you-station-controls/; Spotify, "Discover Hits from Around the World with Spotify's Global Cultures Initiative," *For the Record*, September 28, 2018, https://newsroom.spotify.com/2018-09-28/discover-hits-from-around-the-world-with-spotifys-global-cultures-initiative/; Spotify, "Represent! Celebrate the Diversity of Latin Music for Hispanic Heritage Month," *For the Record*, October 9, 2018, https://newsroom.spotify.com/2018-10-09/represent-celebrate-the-diversity-of-latin-music-for-hispanic-heritage-month/.

39. Seaver, *Computing Taste*, 75.

40. Spotify, "Get Fresh Music Sunup to Sundown with Daylist, Your Ever-Changing Spotify Playlist," *For the Record*, September 12, 2023, https://newsroom.spotify.com/2023-09-12/ever-changing-playlist-daylist-music-for-all-day/.

41. Glenn McDonald, "Every Noise at Once," *Every Noise*, accessed April 16, 2025, https://everynoise.com/engenremap.html.

42. Digital Data Exchange LLC, "Avs:ClassifiedGenre," DDEX Data Dictionary, January 16, 2023, https://service.ddex.net/dd/DD-AVS-CURRENT/dd/avs_ClassifiedGenre.html. This genre list is reproduced directly from the DDEX website. It has been formatted for readability but retains all original grammar and spelling to reflect the source's presentation.

43. Apple does not follow the DDEX standard.

44. A review of the genre list has been launched since my first conversation with DDEX; however, the initial ratio has worsened. In 2024, this was forty-eight for Rock compared to eighteen for Latin.

45. DDEX representative, interview with author, June 2023.

46. "Grove Music Online," accessed April 16, 2025, www.oxfordmusiconline.com/grovemusic/.

47. See for example Greg Anderson's chart of metal genres, which appears in Evelyn Griffin, Anna McClellan, Grace Nugent, and Theo Roe, "Turns Out You Can Judge a Band by Its Album Cover," *The Shield*, September 28, 2020, https://macshieldonline.com/32344/entertainment/turns-out-you-can-judge-a-band-by-its-album-cover/; it is also linked to a Spotify playlist, available at https://open.spotify.com/playlist/oFWYkwwxSklpm8rIqGh1fW?si=CzVEI_1nRcm3TNU13MsQ8Q&nd=1&dlsi=9834410f6e7044e8. Another striking example comes from the taxonomical organization of flamenco, such as the one appearing in Ricardo Molina and Antonio Mairena, *Mundo y Formas del Cante Flamenco* (Madrid: Revista de Occidente, 1963).

48. "Musiio Taxonomy," Musiio Tag, accessed March 22, 2024, https://tag.musiio.com/home/taxonomy. This taxonomy is reproduced directly from the Mussio website. It has been formatted for readability but retains all original grammar and spelling to reflect the source's presentation.

49. Since my conversation with Musiio in 2023, a clarification on the genre tag definition has been added: "Our AI is trained on Bollywood songs backed with Indian instruments, not traditional classical music like Carnatic or Hindustani."

50. Tagging service employee, interview with author, December 2023.

51. Jeremy Wade Morris, "Curation by Code: Infomediaries and the Data Mining of Taste," *European Journal of Cultural Studies* 18, no. 4–5 (2015): 446–63.

52. British Library representative, interview with author, September 2023.

53. Josplay Inc., accessed March 22, 2024, https://josplay.org.

54. British Library representative, interview with author.

55. MarkDDEX, "Early Test of MEAD at Amazon Music and Universal Music Group Leads to an Increase in Active and Passive Streaming Behavior," DDEX, June 15, 2022, https://ddex.net/ddexs-the-big-summit-2022-conference-and-plenary-meeting/.

56. Streaming informant, interview with author, July 2023.

57. Streaming informant, interview with author.

58. See for example Spotify's PR about its Global Cultures Initiative, "Discover Hits from around the World with Spotify's Global Cultures Initiative," *For the Record*, September 28, 2018, https://newsroom.spotify.com/2018-09-28/discover-hits-from-around-the-world-with-spotifys-global

-cultures-initiative/; or the coverage of curatorial efforts in Reggie Ugwu, "Inside the Playlist Factory," *BuzzFeed*, July 13, 2016, www.buzzfeed.com/reggieugwu/the-unsung-heroes-of-the-music-streaming -boom; and Phillip Merrill, "Pandora Goes Curated with Streaming Playlists On-Demand," *Grammy Awards*, October 31, 2017, www.grammy.com/news/pandora-goes-curated-streaming-playlists-demand.

59. Amy Malek, "Clickbait Orientalism and Vintage Iranian Snapshots," *International Journal of Cultural Studies* 24, no. 2 (2020): 266–89.

60. Veronika Muchitsch, "'Genrefluid' Spotify Playlists and Mediations of Genre and Identity in Music Streaming," *IASPM Journal* 13, no. 3 (2023): 48–65; Leight, "Future 25"; Spotify, "Get Fresh Music Sunup to Sundown."

61. Streaming informant, interview with author.

62. "White Supremacist Music Prevalent on Spotify, While Platform Largely Declines to Act," *Anti-Defamation League*, September 22, 2022, www.adl.org/resources/article/white-supremacist -music-prevalent-spotify-while-platform-largely-declines-act.

63. Streaming informant, interview with author.

64. Kassabian, "Would You Like Some World Music with Your Latte?"

65. Ariadna Matamoros-Fernandez, "Platformed Racism: The Mediation and Circulation of an Australian Race-Based Controversy on Twitter, Facebook and YouTube," *Information, Communication and Society* 20, no. 6 (2017): 930–46.

66. Laura Stampler, "Here Are Spotify's First-Ever TV Commercials," *Business Insider*, archived March 27, 2013, at https://web.archive.org/web/20130327175759/www.businessinsider.com/this-is -spotifys-first-ever-commercial-2013-3.

67. Daisy Wyatt, "Tidal Launch: The Most Pretentious Lines from Alicia Keys' Valedictory Speech," *Independent*, March 31, 2015, www.independent.co.uk/arts-entertainment/music/news/tidal-launch -the-most-pretentious-lines-from-alicia-keys-valedictory-speech-a1115441.html.

68. Hanna Appel, Nikhil Anand, and Akhil Gupta, "Introduction: Temporality, Politics, and the Promise of Infrastructure," in *The Promise of Infrastructure*, ed. Appel, Anand, and Gupta (Durham, NC: Duke University Press, 2018), 1–38.

69. Appel, Anand, and Gupta, "Introduction," 26–27.

70. Hesmondhalgh, Campos Valverde, Kaye, and Li, "Digital Platforms and Infrastructure in the Realm of Culture."

71. Appel, Anand, and Gupta, "Introduction," 23.

72. Appel, Anand, and Gupta, "Introduction," 28.

73. Appel, Anand, and Gupta, "Introduction," 24.

74. Morris, "Curation by Code."

13

Streaming into the Metaverse

Jeremy Wade Morris

INTRODUCTION

In May 2022, music streaming giant Spotify made an announcement that differed from its usual updates. Rather than debut a new feature for its music player's interface, detail the acquisition of a burgeoning audio service, or outline an exclusive content partnership with an advertiser or celebrity, Spotify announced it was officially entering the metaverse.[1] Specifically, Spotify was launching a virtual space called "Spotify Island" in Roblox, the popular open-world gaming platform that lets its users create, play, and share game experiences. According to the press release, Spotify Island would be "a paradise of sound where fans and artists from all over the world can hang out and explore a wonderland of sounds, quests, and exclusive merch."[2] Roblox users could visit Spotify Island to play music-based games and experience interacting with the Spotify brand in a way that was different from cuing up a song, building a playlist, or exploring their latest "Discover Weekly." While dozens of big-name media and entertainment brands (e.g., Nike, Adidas, Coca-Cola, Mattel, Disney, Lego, and DreamWorks) had already announced metaverse initiatives by 2022, Spotify was the first major music streaming service to declare a presence in Roblox. In doing so, Spotify became one of a growing number of music industry companies to invest significantly in what continues to be a rather dynamic, uncertain, and speculative space.

Spurred on by pandemic-fueled experiments with alternative concert experiences, livestreamed performances on social media, video game musical events, and other augmented or virtual reality technologies that bridge the physical and the digital, initiatives like Spotify's have coalesced in an emerging industrial buzz phrase: *music in the metaverse.* A complicated and ever-expanding umbrella metaphor, music in the metaverse stands in for all kinds of digital sonic experiments using virtual worlds. It encompasses virtual spaces like Meta's Horizon Worlds that are typically visited with VR headsets or goggles, platforms like Roblox, Fortnite, Sandbox, and other gaming or game-adjacent services that don't necessarily require virtual reality accessories, as well as digital avatars that appear in virtual spaces and on social media platforms, like the virtual idols on the roster of Chinese music label Modern Sky.

In this chapter, I explore various metaverse offerings and how they have shaped a variety of investments from different corners of the music industries. Of particular interest is the way that music streaming services like Spotify, Tidal, or Tencent's QQ Music have envisioned a place for themselves in the metaverse and how other music-related stakeholders have turned to the metaverse as another space for the presentation, distribution, and circulation of musical commodities. Using an analysis of the industrial discourse around the concept of the "metaverse" in trade publications and websites, followed by case studies of several experiments in the musical metaverse by three streaming platforms (Spotify, Tidal, QQ) and other music labels, this chapter considers how music in the metaverse provides models of commodification that rely on speculative experiences, properties, and commodities. Building on research on live and virtual music[3] and on the convergence of the music and technology industries,[4] this chapter examines how such convergence creates additional gatekeepers and spaces for the circulation of music commodities. These intermediaries alter the practices of musicians, fans, and the traditional gatekeepers in the music industries, even as they reinforce some of the structural advantages held by the latter.

The idea of the metaverse as a technological trend attracted considerable media and music industry interest from 2020 to 2023, but this waned in 2024 as attention turned to generative artificial intelligence. Nevertheless, the metaphor of music in the metaverse has already created new experiences with music and new forms of interaction, creating valuable resources and partnerships that are likely to persist even as the metaverse splinters into a distributed network of self-contained ecosystems. Moreover, these investments and partnerships are global in nature, with companies in Asia, Southeast Asia, and Africa experimenting with building infrastructure to support future musical experiences and commodities for their regions. As these various visions of the metaverse emerge, the rush to stake out land, space, and experiences presents new opportunities for musicians to connect with global audiences but also raises questions about what music will look and sound like in a world where traditional music streaming platforms are just one among many ways music is distributed.

LIVE FROM THE METAVERSE

Before diving into streaming platforms' extensions into the musical metaverse, it is useful to trace how music's presence in these virtual spaces emerged, as well as the relationship between virtual and live concerts. We tend to think of "live" music, as Fabian Holt argues, as performances that "involve physical co-presence in a singular time and space."[5] From the biggest concert arenas to local dive bars or street corners, there are countless spaces where live music performances take place every day. As the globally successful tours of Taylor Swift and Beyoncé in the early 2020s demonstrated, live concerts are a significant component of the music industries as well as an economic driver of related industries.[6] This is partly why the lockdowns associated with the global COVID-19 pandemic took such a significant toll on the music industries. Live events and concerts were upended around the world, leading to a precipitous drop in worldwide revenues for the global live music industry in 2020, which continued well into 2021.[7] In response, artists, concert promoters, labels, and fans turned to a wide range of technologies like livestreaming, video game-based concerts, social media watch parties, and other kinds of virtual performances for their "live" events.[8] Even as the popularity of in-person concerts has bounced back—with record-breaking growth in both 2022 and 2023[9]—many of the virtual technological add-ons from the pandemic have persisted.

To be clear, there were certainly virtual concerts, digital meet and greets, and avatar- or hologram-based musicians well before the pandemic.[10] Particularly in places like China and Japan, virtual performances, digital avatar celebrities, and other forms of live virtual events had been happening for years.[11] There were also previous instances of virtual music performances—be it bands like U2 performing in Second Life,[12] virtual avatars like Hatsune Miku,[14] hybrid art and music projects like Gorillaz,[13] or other fusions of music and gaming technology[15]—but those all seemed relatively disconnected, at least industrially. They were isolated initiatives or experiments that did not necessarily signal wider uptake of particular technologies or platforms. In other words, while virtual music events were not exactly marginal when COVID-19 hit, the full shutdown of large public events brought a new level of urgency and visibility to technologies that facilitated online and virtual musical experiences.

It is here, perhaps, where the concept of "music in the metaverse" matters most: it provides an anchoring term that connects a vast set of otherwise disconnected technologies, services, and initiatives that emerged in the years immediately prior to and during the pandemic. The term "metaverse" originally comes from Neal Stephenson's 1990s science fiction novel *Snow Crash*,[16] though its popularity in tech and music industry discourse can be attributed to Facebook's parent company Meta. Looking to expand beyond its web-based social networking site, Facebook renamed itself Meta in 2021. As part of the rebrand, Meta launched a suite of technologies centered around its Oculus virtual reality headsets and its social virtual worlds—3D-generated spaces like Horizon Worlds and Oculus Venues—where users can create avatars and explore virtual environments to socialize, work, communicate, and play. Meta's metaverse had obvious applications for games, social networking, office meetings, and other corporate purposes, but its early launch campaigns also touted virtual concerts and the opportunity to meet musicians and attend virtual VIP events as key features of the metaverse.

The metaverse, for Meta, represented a virtual platform for commercial and social interactions, all fueled through new technology that could be deployed around the world. Meta even struck up partnerships with India's IT Ministry and Tencent in China to spur development of metaverse software and hardware in those regions.[17]

Putting Meta's public relations aside, however, the metaverse is more than just one company's iteration of it. Some metaverse evangelists argue that it is better understood as "the sum total of all publicly accessible virtual worlds . . . that are connected on an open global network, controlled by none and accessible to all."[18] Some information and computing scholars also argue we should look beyond Meta's narrow definition and understand the metaverse more as a "paraverse," which they describe as "an interconnected web of ubiquitous virtual worlds partly overlapping with and enhancing the physical world [that] enable users . . . to experience and consume user-generated content in an immersive, scalable, synchronous, and persistent environment."[19] Instead of one company's private virtual social network, the paraverse recognizes the "global set of disconnected virtual worlds and platforms working independently to advance virtual world technology and culture."[20] While some of these conceptions of the metaverse may be a bit optimistic—recalling the virtual commons arguments that accompanied to arrival of cyberspace—the idea that one company from one particular geographical region might own or control the metaverse in its entirety, reduces the reality of current metaverse developments. The concept of "the metaverse," in other words, is a bit of a misnomer, implying an illusory cohesion to the disparate services and technologies that provide access to a range of virtual experiences. Meta may control a particular experience in, or a particular form of access to, the metaverse, but many other companies, services, and regions are also creating metaverse experiences of their own.

A host of companies and platforms, including Roblox, Fortnite, Minecraft, Sandbox, Decentraland, PartyNite, and Microsoft's Mesh, claim to provide metaverse experiences for their users.[21] In China, major companies like Tencent have been investing in a variety of metaverse initiatives, and established media brands like Sing!China have created a metaverse (Sing!Meta) for their fans.[22] Companies in other regions, like Africarare with its Ubuntuland metaverse for Africa and Metastar Media with its Artistverse in India,[23] are rushing to be the premiere destination for users to experience more culturally specific virtual content and expand the geographical reach and character of the metaverse. Limiting our view of the metaverse to Meta or Euro-American tech and gaming companies also prevents us from considering the regional and national features emerging in various iterations of the metaverse.

What the label "metaverse" does provide, then, is a global, visible, industrial signpost that helps all kinds of organizations direct investment and attention toward something nameable and recognizable, even if there's little precision currently around what the metaverse actually is. This is certainly true for the musical metaverse, which is very much still in an "embryonal state" and "constantly evolving[,] and different musical stakeholders are enriching its meaning in their own ways."[24] The label has already helped a number of companies (e.g., virtual concert platforms, open-world multiuser environments, 3D video capture companies, and digital avatar creation companies) flourish thanks to investment from music industry entities. All three major global music conglomerates (i.e., Sony, Warner Music Group, and Universal Music Group) have invested in efforts to gain a digital foothold in these emerging spaces. From putting on shows in dedicated concert platforms like Wave to forging partnerships for in-game events and merchandise with Roblox and Minecraft to creating experiences in metaverse platforms like Sandbox and Horizon Worlds, major labels spent much of the early 2020s rushing to invest in metaverse-related companies or acquire them.[25] Many major labels and other entertainment companies like Tencent, in 2023, had "signed" some kind of virtual artist—that is, artists who release music and social media content solely as avatars.[26] The MTV Video Music Awards even added a "Best Metaverse Performance" category in 2022 to acknowledge the number of big-name concerts taking place on these platforms.

Despite this rush of activity, the metaverse remains a precarious industrial space. Meta has publicly backed away from and scaled down investment in some of its metaverse plans.[27] ByteDance (owners of TikTok) and Tencent have also restructured their virtual reality operations, affecting their plans for metaverse technology developments.[28] As interest (and investment) in artificial intelligence has surged among tech companies around the world, the metaverse is no longer the shiny new technological trend. Still, tech companies like Meta, Apple, Tencent, and the African telecom giant MTN Group, along with India's IT ministry, continue to invest in the metaverse, and gaming platforms and entertainment companies like Roblox, Fortnite, Minecraft, and Sing!China are still pursuing metaverse partnerships and initiatives.

STREAMING IN THE METAVERSE

If one of the larger goals of this volume is to understand how music streaming platforms operate in terms of their business models, functions, and interfaces in different nations and regions of the world, then a focus on what streaming services are doing within this embryonic metaverse adds virtual spaces to the geographical scope of the analysis. Both streaming services and virtual spaces are often treated as placeless, not bound by physical borders or traditional geographic divisions. But just as streaming companies create offices and headquarters in different regions and manage their rights and licenses on local levels, virtual spaces are similarly rooted in the features and characteristics of the places from which they emerge. Ubuntuland, for example, is very much a project that aims to carve out a space that digitally unites Africa's many countries and presents a virtual space where their continent's history and culture can be presented. The metaverse can both represent real-world spaces as well as reflect the infrastructures and tech industries of particular countries and regions.

Academic research on streaming services is plentiful, as the introduction to this volume attests. Research has examined streaming in relation to algorithmic personalization,[29] artist remuneration,[30] playlists,[31] industrial effects,[32] infrastructures,[33] political economic arrangements,[34] rhetorics of democratization,[35] and the features and workings of specific streaming services[36] (though YouTube has received considerably less focus as a music platform, despite its increasingly important role as site of music consumption).[37] Beyond streaming platforms, researchers have also examined how livestreaming as a practice from gaming[38] has been ported to music hobbyists and performers,[39] and there's a growing body of scholarship on music in virtual spaces, as noted above.

Streaming, as an academic and technical concept, is both a description of a class of services that defines the modern distribution of music (e.g., MSPs) and a broader term for the transmission of all kinds of digital data (including music and video). Streaming music in the metaverse, then, can describe a number of overlapping practices, potentially encompassing a livestream on Instagram of a musician performing in their home studio, a virtual concert in Roblox, or an interactive chat between a fan and their favorite band through a virtual reality headset.

At the moment, the dominant image of music in the metaverse is one that involves an artist using virtual or augmented reality technologies to provide an interactive concert in a digitally created environment. In their most elaborate forms, these concerts feature avatars of the musician(s) in virtual spaces and invite viewers to attend—also as avatars—through virtual reality goggles or web-based platforms. Some custom VR concert platforms, like the one the tech start-up Wave created for Justin Bieber's metaverse concert in 2021, include a series of motion capture sensors that track an artist's live performance in real time. These sensors transmit data that enables the digital avatar in the virtual concert space to mimic the artist's movements as they perform remotely in a green-screen studio.[40] Other performances, like Travis Scott's concert in Fortnite or Lil Nas X's show in Roblox, rely on prerecorded footage that "streams" at a specific point in time. While it may feel live in the game, as users can interact with different views or perspectives of the show, the footage and performances themselves are not actually taking place in real time. Although attendance figures for these shows are hard to track, they reportedly draw between five million and fifty million viewers, yielding multimillion-dollar revenues from tickets and digital merchandise, depending on the artist.[41] Indian pop star Daler Mehndi—who performed India's "first" metaverse concert in 2022—reportedly drew twenty million viewers worldwide. Nigerian rapper Fecko, along with six other artists from the TV talent show *The Mic: Africa*, performed in Africa's first metaverse concert in a production that spanned units in Dubai, Accra, and several different cities in Argentina.[42]

These metaverse concerts, though, are just the most prominent and visible experiments with music in the metaverse. Beyond these lie a series of less spectacular but equally important initiatives that explore how the metaverse might serve as an additional distribution tool for both streaming music companies and global and regional artists and labels looking to pursue models beyond current streaming platforms. I now turn to consider how music streaming services, specifically, are trying to extend the experience of their platforms into virtual spaces, and to assess what this means for the presentation, commodification, and experience of music. I also explore how these and other musical initiatives in various metaverse platforms raise questions about various regional tensions and the infrastructures that define both the metaverse and "music streaming" more broadly.

SONIC AND BRAND EXPERIENCES
ON SPOTIFY ISLAND

Spotify's foray into the metaverse is designed around its larger brand goal of immersing users "in a world of audio no matter how or where they're listening."[43] For Spotify, this typically means providing streaming music via their mobile, desktop, tablet, auto, or television app interfaces, but the metaverse gives Spotify other avenues through which to connect with users. Spotify Island, bathed in the company's signature green hues, offers users the ability to navigate their avatars through trees, forest paths, and other landscape features. The island also has a number of recognizable shapes, colors, and icons that recall Spotify's music software's interface, such as the "heart" or "like" icons that can be collected and traded for merchandise. Users can complete quests for points, which places them on an in-island leaderboard called "The Charts." There is a giant screen in the middle of the island where musical celebrities show up periodically, and there are features on the island that, when activated, produce music. Jumping on a series of musical notes that reside on the broad leaf of a tall plant, for example, allows users to make basic beats. These virtual beat-maker stations are made possible through Spotify's online digital audio workstation, Soundtrap. Users can share their in-game creations publicly with other users. Spotify's press announcement also teased access to exclusive musical content from artists and in-game-only artist merchandise, with a portion of the sales of the latter going back to the artists.[44]

Spotify Island was also designed as a portal to other music-themed worlds. Spotify has launched two so far: K-Park and Planet Hip Hop. K-Park launched shortly after Spotify Island and served as a K-pop-themed world. The company promised it was "the first of a variety of themed islands . . . and the first in a long line of opportunities for artists and fans to connect in the digital world."[45] Drawing its aesthetic palette from bright and colorful K-pop videos, Spotify hoped K-Park would be "an audio destination" for both fans and artists. The choice to feature K-pop seems based on Spotify's own data about the growing importance of the genre not only in the United States but also globally, where K-Pop streams have increased 107 percent and 230 percent, respectively, since 2018. K-Park was targeted to fans in countries that were the top of the list for K-Pop streams (i.e., the United States, Indonesia, the Philippines, Japan, Mexico, Brazil, Malaysia, Canada, Singapore, and the United Kingdom).[46] Spotify further established partnerships with specific K-pop artists like Stray Kids and Sunmi, allowing fans to buy digital merch and interact with digital avatars of the artists as they completed various missions and quests. Spotify's second musical world, Planet Hip Hop, also touted the global reach of the genre (with forty-four billion average monthly streams), but its partnership was limited to the US-based artist Doechii.[47]

Spotify also brought additional features of its software player to Roblox. In 2022, in conjunction with their increasingly popular end-of-year musical wrap up event known as Spotify Wrapped, the company debuted "WonderWrapped" on Roblox. Built on the same principle as K-Park and Planet Hip Hop, Wonder-Wrapped featured quests, games, virtual merch, and other activities that shared graphical elements with the larger Wrapped campaign. The space also provided a digital photo booth where users could have their picture taken with a dozen different musical artists (including those already involved in Spotify's efforts, like Doechii and Sunmi, but also Tove Lo, Eslabon Armado, Calvin Harris, and more). Unlike metaverse concerts, Spotify Island is not an event-based activity like a live or virtual concert that takes place over a particular period of time. Rather, the world aims to introduce (younger) users to Spotify's brand as a destination for music and audio generation. Data on Spotify's audience for the initiative are limited, but the advertising company that helped create Spotify Island claims the site received over 1.6 billion "global impressions," and Roblox's audience, which skews young, is nearly equally divided between users in North America, Europe, Asia-Pacific, and the rest of the world.[48]

Interestingly, Spotify doesn't actually offer its own streaming music in Spotify Island. There is constantly music in the background, but, like a video game, it's a series of repeating tracks scored for specific spaces in the world. Users either listen to the background score or generate their own music through the virtual beat-making stations. Spotify does curate a playlist on its main music service called "Spotify Island on Roblox," which features about two and a half hours of popular songs across a wide variety of genres and artists. But in Spotify Island, "streaming" is less about the constant flow of music from Spotify's service, and more about creating an interaction between users, the Spotify brand, and its associated technologies (e.g., Soundtrap). There's no direct link to a Spotify player within the environment, so the integration between Spotify's core offering—streaming music—and Roblox's game world does not rely on the same economic and technical mechanisms that underpin streaming (i.e., servers, royalties, etc.). Rather, the metaverse serves as a space to encourage new interactions with music and musical artists in which Spotify has invested. Instead of typical royalty arrangements, artists make direct deals with Spotify and receive exposure and support in return. The deals Spotify strikes with both Roblox and the featured artists are generally not publicly disclosed. Artists do get a percentage of in-game merch sales,[49] as well as other forms of support from Spotify. Doechii's appearance in Hip Hop Planet, for example, was part of a larger promotion of Doechii across the platform, in conjunction with Spotify's global emerging artist program, RADAR.[50] Through RADAR, Spotify partners with artists in countries like Italy, Spain, and Australia and provides support through billboards, platform marketing, and additional social content. The metaverse, then, becomes another channel Spotify uses to help grow and break artists in its various regional markets.

SPECULATIVE ASSETS IN STREAMING SERVICES AND BEYOND

Looking at other popular streaming services shows a different approach to investments in metaverse technologies. While the US-based streaming service Tidal has participated in metaverse performances and events (e.g., sponsoring a performance featuring Charli XCX in Meta's Horizon Venues/Worlds metaverse in 2020 and its "Tidal Rising" virtual concert series in 2022),[51] it hasn't yet created a persistent presence in the metaverse like Spotify Island. Rather, through its partnership with the virtual reality company Sensorium, it has invested in in-world currency that can be translated into other forms of value in the Sensorium Galaxy, a futuristic metaverse that allows users to create digital avatars and attend virtual concerts and events. The project is geared largely towards Euro-American electronic dance music communities and features exclusive virtual concerts by globally recognized DJs and musicians such as Carl Cox, Steve Aoki, Black Coffee, and Charlotte de Witte. Sensorium Galaxy also offers a 24/7 streaming dance party on its website, attended by artificially intelligent avatars and avatars representing everyday users.

In 2020, Tidal purchased US$7 million in tokens issued by Sensorium. SENSO tokens are an Ethereum-based digital currency that drives interactions in Sensorium Galaxy, allowing users to buy custom avatar outfits and pay for concerts and events. At the time of writing, the Sensorium website sells one SENSO token for US$10 and twenty tokens for US$160. Tidal's purchase, then, is a speculative one. The company is hoping to realize value by reselling those tokens to other users or by offering them as a part of a larger promotion involving artists it represents. According to Tidal's COO Lior Tibon, the purchase gave the company the "opportunity to gain exclusive rights for its stellar artist roster to have their shows and music broadcast exclusively within Sensorium's themed virtual entertainment worlds."[52] Hip hop mogul Jay-Z's entertainment company Roc Nation also purchased an undisclosed number of SENSO tokens in 2021, hoping to give its artists a chance to "benefit from global content distribution through Sensorium Galaxy and safeguard ownership rights on all of their digital content."[53]

Similarly, Chinese tech giant Tencent has used its QQ Music service to experiment with metaverse activities. In addition to its TMELAND virtual music carnival in 2021,[54] the company added a "music zone" to the app where users could purchase musically themed avatars, decorations, and accessories in the form of nonfungible tokens (NFTs) to create a personalized space within the app. The Music Zone was meant to be a social networking space supported by a shared interest in music: a virtual avatar–fueled version of MySpace, updated with digital purchasing options (e.g., NFTs). Rather than purchasing music directly, users buy digital versions of music-related commodities, opening another avenue for the sale of traditional merchandise associated with music.

Warner Music Group's 2022 partnership with the metaverse platform Sandbox and Africarare's plans for Ubuntuland show even further investment in musical metaverse infrastructures. Like Spotify Island in Roblox, Warner's deal with Sandbox involved the promise to create a virtual space that was part concert venue and part musical theme park.[55] Unlike traditional metaverse concerts or even Spotify Island, though, the Sandbox allows users to buy "land" or "property" in the virtual space, using SAND tokens, the in-game currency. Users can customize their "property" how they see fit, building any number of digital structures or services into their plot, or renting it out for others to develop. Depending on the market, a plot of land can be worth anywhere from several hundred to tens of thousands of dollars. Warner Music Group's land is considered to be relatively high in value, almost like "beachfront property," because it is near other high-value plots. As WMG's chief digital officer Oana Ruxandra noted, "Our partnership with The Sandbox adds a new layer of possibility in the metaverse, with the ownership of virtual real estate [that provides] persistent, immersive social music experiences that defy real-world limitations and allow our artists and their fans to engage like never before."[56] In 2023, this partnership led to the "Infinite Pulse Land Sale," which involved WMG selling land on its property, located near plots owned by other musical artists like Slipknot, Jamiroquai, and even Elvis Presley's estate.[57] Ubuntuland offers a similar range of digital plots whose price is assigned via periodic private and public auctions. The initiative's website promises that the plots can be used to sell merchandise, host classes, lectures, and workshops, or put on concerts and other kinds of virtual events. It was through its purchase of plots that African telecom company MTN was able to host what it promoted as Africa's first metaverse concert.[58]

This rush for land, and the creation of experiences and events within these spaces echoes what Ulises Mejias and Nick Couldry call data colonialism, though these musical investments remain speculative at the moment.[59] None of the streaming music platforms has started offering streaming services within the metaverse, and many of the music labels or other companies that have set up musical spaces are exploring the sale of digital plots, NFTs, and other digital assets (e.g., skins for users' avatars, accessories, and clothing) based on the artists they represent, rather than direct sales of musical recordings or subscriptions to music services. Three years after Tidal's investment in SENSO tokens, the official launch of the Sensorium Galaxy is still in beta. It has a heavy web presence and several livestream concerts available, though Reddit threads and other online forums question whether development on the initiative has indeed completely stopped.[60] Users report similar difficulties purchasing plots in Ubuntuland.[61]

The ambivalent status of these sites merely underscores the speculative nature of so many of the promises made around the metaverse. The metaverse offers musicians and their labels the chance to be investors and speculators, with the music in these spaces becoming the added value or distinguishing feature of their particular corner of the metaverse. This is not an unfamiliar position for record labels; they have long made bets on which artists they thought might be successful enough to invest in. Now, however, the speculative investing seems to be in the future value of digital commodities like land, accessories, and other assets. Mirroring the more general financialization that has taken place in the music industries—where private equity firms and other investment organizations like BlackRock, JPMorgan Chase, Hipgnosis, and Kohlberg Kravis Roberts bet millions on acquiring publishing catalogs in the hopes of future revenues[62]—the metaverse investments seem similarly dependent on some future ability to realize value from a space, image, or digital object, in addition to the usual process of betting on an artist's music and potential for celebrity. Financial firms buying up music catalogs believe that these assets are undervalued and offer significant potential for long-term exploitation of intellectual property. The same speculative logic seems to be driving the metaverse—and the related technologies on which it relies, such as cryptocurrency, NFTs, and blockchain-enabled commerce—even though there are fewer concrete examples, at least at this point, of how the seemingly infinite possibilities the metaverse may provide artists, distributors, and users will be logistically realized.

STREAMING EXPERIENCES AND MINTING
NEW MUSICAL COMMODITIES

After decades of tumult in terms of revenues from recorded music commodities, revenues from streaming services have stabilized and continue to grow predictably for the majors.[63] The metaverse, in this light, represents an opportunity to experiment beyond streaming for new kinds of commercialization and revenues. The efforts by Spotify, Tidal, and Tencent are examples of how streaming services themselves see the metaverse as a place for musical experiences to further their brands, but the wider push by labels and metaverse platforms to make the metaverse more musical suggests a desire to bet on future technologies and technological infrastructure that positions them as the landowners and commodity producers in these emerging virtual spaces.

Given the emphasis on the purchase and exchange of a variety of commodities within the musical metaverse—be it parcels of land, digital tickets and souvenirs, or custom avatars of your favorite artists—the metaverse environment gives rights holders with a new space that they see as potentially providing infinite possibilities for commodification. The endless digital offerings that could conceivably emerge represent a tantalizing opportunity for an industry that spent the better part of the first decade of this century figuring out how to repackage the recorded music commodity as a sellable thing in light of widespread file-sharing. Taking cues from the gaming platforms with which they now often partner, music companies are using the metaverse to experiment with economic models that mirror the gaming industry—a kind of lab for testing additional revenue streams adjacent to their core product. For gaming, close to 75 percent of the industry's revenues come from engagement after acquiring the product for free (e.g., spending their money on in-game or in-app purchases), while the remaining 25 percent comes from paying for access to the product.[64] The music industries, in the metaverse, are seeking something similar. Music had previously excelled at monetizing the moment of access, and streaming provides one solution to that problem, but there is room to grow in terms of engagement. Through digital assets in the metaverse and plots of land built around musical experiences, futurist executives hope to overcome this challenge by giving fans different forms of access to artists and different modes of engaging with musical communities and fandoms. If streaming prices are, at their core, rooted in deals that emerged nearly fifteen years ago in the risk-averse aftermath of file-sharing, current music industry thinking envisions the metaverse as an additional financial *and* engagement opportunity. As WMG's Oana Ruxandra notes, "We have a world today that I believe is feeling the confines of that risk aversion, and a lot of consumers are looking to expand the way in which they engage with music."[65]

For streaming platforms, then, the metaverse represents a space to extend their brands and introduce other services that may supplement their businesses. They have yet to become the soundtrack of the metaverse through more immediate integration (i.e., a Spotify or QQ Music player interface accessible within the metaverse), but they have recognized the metaverse as a space to court new (young) users and highlight other elements of their services (e.g., Soundtrap, RADAR, Music Zone, etc.). These spaces also offer labels and artists the opportunity to develop new kinds of musical commodities and extend the concept of a "streaming service" to a more diffuse musical experience in a virtual space. Proponents of music in the metaverse assume that instead of merely pressing play on a song in their streaming service, users will want to immerse themselves in musical environments, where concerts, music generation, and interactions with a host of musical commodities will generate engagement, rather than just provide access. These speculative visions will likely generate new streams of revenue for musicians, labels, and metaverse platforms alike; whether they will also reconfigure the idea of music streaming remains as speculative as some of the investments these companies are making.

NOTES

1. Spotify, "Spotify Island Brings New Experiences for Fans and Artists to Roblox," *For the Record*, May 3, 2022, https://newsroom.spotify.com/2022-05-03/spotify-island-brings-new-experiences-for-fans-and-artists-to-roblox/.

2. Spotify, "Spotify Island."

3. Philip Auslander, *Liveness: Performance in a Mediatized Culture* (London: Routledge, 1999); Sheila Whiteley and Shara Rambarran, eds., *The Oxford Handbook of Music and Virtuality* (Oxford: Oxford University Press, 2016).

4. David Hesmondhalgh and Leslie Meier, "What the Digitalisation of Music Tells Us about Capitalism, Culture and the Power of the Information Technology Sector," *Information, Communication and Society* 21, no. 11 (2018): 1555–70; Qian Zhang and Keith Negus, "From Cultural Intermediaries to Platform Adaptors: The Transformation of Music Planning and Artist Acquisition in the Chinese Music Industry," *New Media and Society* (2024), https://doi.org/10.1177/14614448241232086.

5. Fabian Holt, "The Economy of Live Music in the Digital Age," *European Journal of Cultural Studies* 13, no. 2 (2010): 252.

6. Pallavi Gogoi, "From Swiftie Beads to Barbie and Beyoncé, Girls Ran the World Economy in 2023," *NPR*, December 27, 2023, www.npr.org/2023/12/27/1221483930/girl-economy-taylor-swift-beyonce-barbie-2023.

7. Janis Denk, Alexa Burmester, Michael Kandziora, and Michel Clement, "The Impact of COVID-19 on Music Consumption and Music Spending," *PLoS ONE* 17, no. 5 (2022), https://doi.org/10.1371/journal.pone.0267640.

8. Jeremy Wade Morris, "Live at the App: The Economics, Platforms, and Technologies of Livestreamed Music," in *The Sage Handbook of the Digital Media Economy*, ed. Terry Flew, Jennifer Holt, and Julian Thomas (London: Sage, 2023), 260–79.

9. Thania Garcia, "Live Music Logs Record-Setting 2022 as Bad Bunny and Elton John Lead with Booming Stadium Tours," *Variety*, December 13, 2022, https://variety.com/2022/music/news/top-tours-2022-bad-bunny-elton-john-stadiums-1235458652/; Stuart Dredge, "Live Nation Reveals 20% Growth for Concert Attendance in 2023," *Music Ally*, February 23, 2024, https://musically.com/2024/02/23/live-nation-reveals-20-growth-for-concert-attendance-in-2023/.

10. Whiteley and Rambarran, *Music and Virtuality*; Qian Zhang and Keith Negus, "Stages, Platforms, Streams: The Economies and Industries of Live Music after Digitalization," *Popular Music and Society* 44, no. 5 (2021): 539–57.

11. Ka Yan Lam, "The Hatsune Miku Phenomenon: More Than a Virtual J-Pop Diva," *Journal of Popular Culture* 49, no. 5 (2016): 1107–24.

12. Justin Gagen and Nicholas Cook, "Performing Live in Second Life," in Whiteley and Rambarran, 191–209.

13. Shara Rambarran, *Virtual Music: Sound, Music, and Image in the Digital Era* (New York: Bloomsbury, 2021).

14. Lam, "Hatsune Miku."

15. Karen Collins, *Playing with Sound: A Theory of Interacting with Sound and Music in Video Games* (Cambridge, MA: MIT, 2013).

16. Neal Stephenson, *Snow Crash* (1992; repr., New York: Random House Worlds, 2000).

17. Lynn Doan and Edward Ludlow, "Meta Nears Deal to Sell Mixed-Reality Headsets in China with Tencent Partnership," *Bloomberg*, November 10, 2023, www.bloomberg.com/news/articles/2023-11-10/meta-closing-in-on-deal-to-sell-mixed-reality-headsets-in-china; Manish Singh, "Facebook Parent

Meta Launches Startup Accelerator with India's IT Ministry in Metaverse Push," *TechCrunch*, September 13, 2022, https://techcrunch.com/2022/09/13/facebook-parent-meta-launches-startup-accelerator-with-indias-it-ministry-in-metaverse-push/.

18. Tony Parisi, "The Seven Rules of the Metaverse," *Metaverses*, October 23, 2021, https://medium.com/meta-verses/the-seven-rules-of-the-metaverse-7d4e06fa864c.

19. Markus Weinberger quoted in Pat Healy and Hannah Standiford, "Enter the Paraverse: Challenging Assumptions of Live Music in the Metaverse," *European Conference on Games Based Learning* 17, no. 1 (2023): 252.

20. Healy and Standiford, "Enter the Paraverse," 252.

21. Jacob Kastrenakes, "Roblox Goes Public So That It Can Build a Bigger Metaverse," *The Verge*, November 19, 2020, www.theverge.com/2020/11/19/21578491/roblox-ipo-announced-dau-increase-pandemic; Paul Tassi, "Fortnite's Epic Games Makes a Metaverse Investment to Scale Up Even Further," *Forbes*, September 23, 2022, www.forbes.com/sites/paultassi/2022/09/23/fortnites-epic-games-makes-a-metaverse-investment-to-scale-up-even-further/.

22. "Sing!China Metaverse Welcomes 100 Million Viewers and Fans, Unlocking a New Interactive Web3 Experience," *European Business Magazine*, July 14, 2023, https://europeanbusinessmagazine.com/media-outreach/singchina-metaverse-welcomes-100-million-viewers-and-fans-unlocking-a-new-interactive-web3-experience/; "Tencent's QQ Music Is Testing Metaverse Social Feature: Report," *TechNode*, July 5, 2022, https://technode.com/2022/07/05/tencents-qq-music-is-testing-metaverse-social-feature-report/.

23. Sneha Bhura, "How Indian Musicians Are Drawing Their Followers into the Metaverse," *Times of India*, April 23, 2024, https://timesofindia.indiatimes.com/india/how-indian-musicians-are-drawing-their-followers-into-the-metaverse/articleshow/109536023.cms; William Brederode, "Africa's First Metaverse Wants to Bring a Billion People Together in a Mixed Reality World," *News24*, August 28, 2023, www.news24.com/news24/tech-and-trends/africas-first-metaverse-wants-to-bring-a-billion-people-together-in-a-mixed-reality-world-20230828.

24. Luca Turchet, "Musical Metaverse: Vision, Opportunities, and Challenges," *Personal and Ubiquitous Computing* 27, no. 5 (2023): 1812.

25. Jamie MacEwan, Alice Enders, Niamh Burns, and Sholto Cocris, "Welcome to the Metaverse: Music in the Metaverse," Enders Analysis, October 4, 2022, www.endersanalysis.com/reports/welcome-metaverse-music-metaverse.

26. Daniel Tencer, "Meet the Virtual Artists Backed by Some of the World's Biggest Entertainment Companies," *Music Business Worldwide*, May 22, 2023, www.musicbusinessworldwide.com/meet-the-virtual-artists-backed-by-some-of-the-worlds-biggest-entertainment-companies1/.

27. Ananya Bhattacharya, "Meta's 'Year of Efficiency' Means Job Cuts, Less Metaverse, and More Generative AI," *Quartz*, March 7, 2023, https://qz.com/meta-layoffs-2023-jobs-metaverse-ai-1850196575.

28. Nina Xiang, "Metaverse No More? ByteDance and Tencent Scale Back VR Ambitions," *Forbes*, November 17, 2023, www.forbes.com/sites/ninaxiang/2023/11/17/metaverse-no-more-bytedance-and-tencent-scale-back-vr-ambitions/.

29. Robert Prey, "Nothing Personal: Algorithmic Individuation on Music Streaming Platforms," *Media, Culture and Society* 40, no. 7 (2018): 1086–1100.

30. David Hesmondhalgh, "Is Music Streaming Bad for Musicians? Problems of Evidence and Argument," *New Media and Society* 23, no. 12 (2021): 3593–3615.

31. Anja Nylund Hagen, "The Playlist Experience: Personal Playlists in Music Streaming Services," *Popular Music and Society* 38, no. 5 (2015): 625–45.

32. Keith Negus, "From Creator to Data: The Post-Record Music Industry and the Digital Conglomerates," *Media, Culture and Society* 41, no. 3 (2019): 367–84.

33. Paolo Magaudda, "The Future of Digital Music Infrastructures: Expectations and Promises of the Blockchain 'Revolution': Politics, Economy, Culture and Technology," in *Popular Music in the Post-Digital Age: Politics, Economy, Culture and Technology*, ed. Ewa Mazierska, Les Gillon, and Tony Rigg (London: Bloomsbury, 2019), 51–68.

34. Eric Drott, *Streaming Music, Streaming Capital* (Durham, NC: Duke University Press, 2024).

35. Raphaël Nowak and Benjamin A. Morgan, "New Model, Same Old Stories? Reproducing Narratives of Democratization in Music Streaming Debates," in *New Model, Same Old Stories? Reproducing Narratives of Democratization in Music Streaming Debates*, ed. Marko Kölbl and Fritz Trümpi (Vienna: mdwPress, 2021), 61–84.

36. Maria Eriksson, Rasmus Fleischer, Anna Johansson, Pelle Snickars, and Patrick Vonderau, *Spotify Teardown: Inside the Black Box of Streaming Music* (Cambridge, MA: MIT Press, 2019).

37. Jean-Samuel Beuscart, Samuel Coavoux, and Jean-Baptiste Garrocq, "Listening to Music Videos on YouTube: Digital Consumption Practices and the Environmental Impact of Streaming," *Journal of Consumer Culture* 23, no. 3 (2023): 654–71.

38. T. L. Taylor, *Watch Me Play: Twitch and the Rise of Game Live Streaming* (Princeton, NJ: Princeton University Press, 2018).

39. Anne Danielsen and Yngvar Kjus, "The Mediated Festival: Live Music as Trigger of Streaming and Social Media Engagement," *Convergence* 25, no. 4 (2019): 714–34; Mark Thomas, "Digital Performances: Live Streaming Music and the Documentation of the Creative Process," in *The Future of Live Music*, ed. Ewa Mazierska, Les Gillon, and Tony Rigg (New York: Bloomsbury, 2020), 83–97; Zhang and Negus, "Stages, Platforms, Streams."

40. Lauren Huff, "Justin Bieber Is Going into the Metaverse for a Live Virtual Concert—Watch the Trailer," *Entertainment Weekly*, November 17, 2021, https://ew.com/music/justin-bieber-metaverse-virtual-concert-trailer/.

41. Justin Patel, "Top 10 Most Popular Metaverse Concerts," *Metaverse Marcom*, November 18, 2022, www.metaversemarcom.io/post/top-10-most-popular-metaverse-concerts.

42. "Take Back the Mic Studios Makes History with Live-Action Performances in African Metaverse," *Black Enterprise: Wealth for Life*, December 26, 2022, www.blackenterprise.com/take-back-the-mic-tbtm-studios-makes-history-with-live-action-performances-in-the-african-metaverse/; Raisa Sarkar, "20 Million Viewers Tuned in to Watch India's First-Ever Metaverse Concert Led by Celebrated Indian Pop Icon Daler Mehndi," *Digital Studio: Broadcasting and Production in India*, February 22, 2022, www.digitalstudioindia.com/events/20-million-viewers-tuned-in-to-watch-indias-first-ever-metaverse-concert-led-by-celebrated-indian-pop-icon-daler-mehndi.

43. Spotify, "Spotify Island."

44. Billy Steele, "Spotify Enters the Metaverse with Spotify Island on Roblox," *Engadget*, May 3, 2022, www.engadget.com/spotify-island-roblox-metaverse-110035809.html.

45. Spotify, "Take a Tour of Spotify Island's New K-Park on Roblox," *For the Record*, May 24, 2022, https://pr-newsroom-wp.appspot.com/2022-05-24/take-a-tour-of-spotify-islands-new-k-park-on-roblox/.

46. Spotify, "Take a Tour."

47. Spotify, "Here's Your First Look at Planet Hip-Hop, Spotify Island's Latest Experience on Roblox," *For the Record*, September 21, 2022, https://newsroom.spotify.com/2022-09-21/heres-your-first-look-at-planet-hip-hop-spotify-islands-latest-experience-on-roblox/.

48. "Spotify Island on Roblox," Spotify, accessed May 4, 2024, https://ourcasestudy.byspotify.com/roblox; Ariana Newhouse, "Everything You Need to Know on Roblox Stats and Spending," *The Shelf*, March 21, 2024, www.theshelf.com/the-blog/roblox-stats-and-spending/.

49. Steele, "Spotify Enters the Metaverse."

50. Spotify, "Spotify's RADAR Program Brings a Fresh Look and New Stories from Global Artists, Songwriters, and Podcasters," *For the Record*, October 5, 2023, https://newsroom.spotify.com/2023-10-05/radar-artists-podcasters-songwriters-new-look/.

51. Stuart Dredge, "Meta and Tidal Team up for New Series of Virtual Concerts," *Music Ally*, October 26, 2022, http://musically.com/2022/10/26/meta-and-tidal-virtual-concerts/.

52. Stuart Dredge, "Tidal Makes Social VR Move with $7m Purchase of Sensorium Tokens," *Music Ally*, August 27, 2020, http://musically.com/2020/08/27/tidal-makes-social-vr-move-with-7m-investment-in-sensorium/.

53. Brian Kean, chief communications officer at Sensorium, quoted in Murray Stassen, "Jay Z's Roc Nation Makes VR Move with Purchase of SENSO Tokens from $100m-Backed Sensorium Corporation," *Music Business Worldwide*, July 7, 2021, www.musicbusinessworldwide.com/jay-zs-roc-nation -makes-vr-move-with-purchase-of-senso-tokens-from-100m-backed-sensorium-corporation/.

54. Julienna Law, "Adidas Throws a Music Party in China's Metaverse," *Jing Daily*, May 24, 2022, https://jingdaily.com/posts/adidas-tencent-ozworld-metaverse-festival; "Tencent's QQ Music Tips into Virtual Community," PingWest, July 5, 2022, https://en.pingwest.com/w/10454.

55. Adi Robertson, "Warner Music Group Is Launching a Metaverse Concert Hall Where You Can Pay to Be Its Neighbor," *The Verge*, January 27, 2022, www.theverge.com/2022/1/27/22904382/warner -music-group-the-sandbox-virtual-real-estate-sale-concert-venue.

56. Warner Music Group. "The Sandbox Partners with Warner Music Group." *Warner Music Group News Blog*, January 27, 2022, https://www.wmg.com/news/sandbox-partners-warner-music -group-create-music-themed-world-metaverse-36116.

57. "Introducing Infinite Pulse: Unleashing the Power of Music and Blockchain in Musicverse," *The Sandbox*, July 27, 2023, https://medium.com/sandbox-game/introducing-infinite-pulse-unleashing-the -power-of-music-and-blockchain-in-musicverse-dea8de31144e.

58. Brederode, "Africa's First Metaverse."

59. Ulises A. Mejias and Nick Couldry, *Data Grab: The New Colonialism of Big Tech and How to Fight Back* (Chicago: University of Chicago Press, 2024).

60. "Sensorium Galaxy," Reddit, accessed May 2, 2024, www.reddit.com/r/SensoriumGalaxy/.

61. Jan Vermeulen, "We Tried to Buy Land in Ubuntuland's Metaverse—but It Was Impossible," *MyBroadband*, April 22, 2022, https://mybroadband.co.za/news/cryptocurrency/442064-we-tried-to -buy-plots-in-ubuntulands-metaverse-but-it-was-impossible.html.

62. Tim Ingham, "Music Catalogs Are Selling for Serious Cash. Now Wall Street Wants In," *Rolling Stone*, January 13, 2021, www.rollingstone.com/pro/features/music-catalogs-are-selling-for-serious -cash-now-wall-street-wants-in-on-it-1113766/.

63. Bill Rosenblatt, "Music Industry Annual Reports Show Stable Growth and YouTube Strength," *Forbes*, March 30, 2024, www.forbes.com/sites/billrosenblatt/2024/03/30/music-industry-annual -reports-show-stable-growth-and-youtube-strength/.

64. Stuart Dredge, "WMG's Oana Ruxandra on Music's Evolution: 'Conservatism Is Out the Door!,'" *Music Ally*, January 16, 2023, http://musically.com/2023/01/16/wmg-oana-ruxandra-music-evolution/.

65. Dredge, "WMG's Oana Ruxandra on Music's Evolution."

Afterword

Music Streaming and Throwing Stones

Yiu Fai Chow

When Dave invited me to write an afterword for this collection of essays, my immediate reaction was, "Me?" Of course, I was honored and flattered. David Hesmondhalgh is the author of many important and inspirational works, from which I have drawn insights for my own research and teaching on creative practices. I must add that they have also contributed to my self-understanding and well-being as a creative practitioner. I have been writing lyrics for Chinese-language popular music since 1989, long before I started my academic life. When, more recently, I learned that Dave was embarking on a new project on music streaming platforms, I looked forward to an eventual output like the one you are reading. As a researcher, practitioner, and user (or should I say consumer, listener, or audience member?), I have been sharply aware of the impact of music streaming, and critically curious about it. I recall the first time I was commissioned to write lyrics for a song to be released by one of the platforms in mainland China. The producer conveyed the platform's request for full ownership, which was at odds with the common industry model of rights distribution. The unusual request is enabled by the unusual position—or power—that such platforms hold or are perceived to hold. As Hesmondhalgh has noted, "The massive new role of Music Streaming Platforms (MSPs) in musical consumption means that they increasingly operate as the core of the music industries and of the everyday experience of recorded music."[1]

I want to know more; I want to read more. But to write an afterword? Someone who is not an expert in MSPs? Someone who is not exactly following scholarly development in the field of MSP studies?[2] In the end, I said yes to Dave. How could I not? I realized that while my academic connection to Hesmondhalgh's current project may be tenuous, my personal resonance with Dave persuades me in a way that does not feel like persuasion. I cited his works when I was doing my PhD, and I connected with Dave personally when I became a professional academic. I'll skip further compliments and confine myself to admitting that I decided to let citation and connection go hand in hand. I said yes. And I have chosen to present this "yes" in such an elaborate manner not only because I want to premise this afterword with a sort of imposter syndrome camouflaged as a caveat, but also because I want to highlight certain keywords in my own practice: personal, contingent, affective. They will serve as a linchpin, helping me to respond to this rich array of contributions.

For the rest of this afterword, bear in mind my speaking position—first, an intersection between creative practice studies, gender studies, and cultural studies; second, an intersection between me as a researcher and me as a practitioner. Let me start with cultural studies. In 2010, Lawrence Grossberg published an influential article in which he posited the political responsibilities of cultural studies as follows: "The project of cultural studies is to tell better stories about what is going on, and to begin to enable imagining new possibilities for a future that can be reached from the present—one more humane and just than that promised by the trajectories we find ourselves in."[3] This volume and the project that lies behind it are part and parcel of this cultural studies project—this attempt to present and represent the transformations we are experiencing, in both music and the world at large. So much has been going on, and going on so fast, that one cannot help but wonder, indeed: what is going on? Taken together, the contributions are grinding the fast-changing world to a halt, or at least putting it on hold, to sound out a gentle but forceful rejoinder: this is what is going on—all of this.

My understanding of Grossberg's "better" stories, and my immediate gratitude to the authors gathered here, is precisely the collection of "all of this." While cultural studies scholars continue to have their take on what constitutes better stories,[4] I see better stories primarily as more stories—the more, the better—as multiplicity wedging open simplicity and, thus, inevitability; as a collective and connective manner of recognizing lives lived differentially and inequitably and, thus, recognizing difference and inequity itself. More stories function to ultimately foreground the inadequacies of master narratives. In terms of scholarship on technology or technological developments, there has been a well-documented tradition of binary thinking: utopian-dystopian, often built on what Raymond Williams called "technological determinism" and its apparently ceaseless afterlives.[5] Half a century ago, Marshall McLuhan offered a master narrative of the "global village" where the speed of new—new at that time—electronic communication would usher humanity into harmonization, connection, and homogeneity.[6] One can easily find the latest manifestation of such binarism surrounding thinking on big data and AI: humankind's gravest threat or best advance, the demise or rebirth of humanity, and so forth.[7] This book steers us away from the seduction of master narratives and their answers and solutions. These narratives are not always or necessarily incorrect, but they should not serve to eclipse, configure, or dictate what is going on, nor should they confine the proliferation of possibilities.

THE MORE, THE BETTER

David Hesmondhalgh articulates the central concern of his project as follows: "The main focus of this volume . . . is to understand the effects of streaming on the circulation of music across nations and continents, and what those effects tell us about power, justice, and inequality in music culture."[8] Extrapolated into geopolitical terms, this concern is grounded in the already dominant position of Western music production and consumption, as well as representation and imagination, and its possible imbrication with the latest technological development of MSPs at the continued expense of the Global South—or quite the opposite? Instead of delivering a verdict on "digital colonialism" or "platform imperialism," this volume presents a multiplicity of experiences that, taken together, speak against such master narratives as technological determinism while at the same time offering individual, locally inflected instances, compelling us to understand what is going on globally in an individual, locally inflected manner. The stories collected here do not allow an optimistic scenario of increasing equality and democratization among music practitioners around the world, but they also foreground what Hesmondhalgh calls, in his introductory chapter in this volume, "a new musical multipolarity"—in other words, multiple, complex, more.

That we are updated with experiences from the Global South is, of course, a feat in itself. The collection includes, refreshingly, chapters dedicated to China, Egypt, India, Japan, Kenya, Mexico, and Turkey. Even in the so-called Global North, the volume brings in more than the usual suspects—that is, beyond or in juxtaposition with the United States and the United Kingdom. There are fine contributions from Hungary (by Emília Barna), Italy (by Francesco D'Amato), and a comparative piece connecting the United States and the United Kingdom with Norway, Germany, and France (by Raquel Campos Valverde). One chapter (by Jeremy Wade Morris) is "extraterritorial," as it covers the metaverse. As I noted in a coauthored book, despite decades of postcolonial realities and studies, knowledge production in popular music remains mired in power imbalances, not unlike the production of popular music itself. There is a need to do more "scholarship that seeks to de-Westernize popular music studies, a field of knowledge production persistently dominated by Anglo-Saxon experience and publications."[9]

This supplement is always already an enrichment—not only in terms of geographical coverage or in interrogating the domination of the Global North, but above all, in highlighting a diversity of experiences, cultures, and logics, and thus disparities, disjunctures, and ultimately dissent. Even when we admit the globalization of MSPs and their modus operandi with its concomitant ideology, we also read from the scenarios garnered around the world the possibilities of doing otherwise. Such possibilities may not last, but they are there. After all, what is lasting when we also read the genealogies of the technological develop-ments prior to MSPs? Various authors have offered historical accounts in con-nection to contributions from the global scenarios—vinyl, cassette tapes, CDs, MTV—and narrated in that sense: nothing lasts or must persist, inevitably, as such. Noriko Manabe's chapter seems to me the most illustrative of this dissent toward the inevitable. It opens with figures probably astonishing for readers not familiar with the Japanese music market. "According to the International Fed-eration of the Phonographic Industry (IFPI), streaming accounted for 67 per-cent of global music revenues in 2023, while physical sales have dwindled to 18 percent; in Japan, these figures are flipped, with physical sales still accounting for 65 percent of revenues and streaming accounting for 31 percent—roughly equivalent to the US market eight years prior, in 2015."[10] While some streaming experts predict change, Manabe's analysis shows a corporate-cultural nexus that says, at the very least, that things can be different.

We must heed the local and the regional, as well as the global, as inflected in the local and the regional. In this process, one very specific contribution to this vol-ume is as conceptual as it is factual. It introduces to us the richness of local and regional music. By analyzing how they engage with global MSPs, a host of authors have enlightened readers—at least, they have enlightened me—about music, artists, and platforms quite unheard of outside local and regional contexts in a music world dominated by the United States and the United Kingdom, and MSPs such as Spotify. Conceptually, this is quite simply a corollary of what I wrote above. Here, however, I want to highlight the factual, or perhaps informative, dimension—namely, my gratitude to the authors for bringing me and fellow readers to the local and regional scenario. For instance, from Rodrigo Gómez, Ignacio Gallego, and Argelia Muñoz-Larroa, I learn of the Mexican genre corridos tumbados; and from Aditya Lal of non-Bollywood music and artists such as King, Arijit Singh, and Diljit Dosanjh. From Darci Sprengel, I understand that Anghami may be the platform of choice for music in the Middle East and North Africa / Southwest Asia and North Africa region. This is not big data, but small data—at once personal and political.

TECHNOLOGY, CULTURE

Privileging small data is, of course, my small way of troubling the concern with MSPs or platforms in general; their concern is *big* data. An episode relayed by Spren-gel in her chapter is telling. During a discussion on the financial benefits the internet brings to independent musicians, Rami Zeidan, product director of Anghami, the "local" MSP I just mentioned, evaded the ethics of fairness and spoke the language of data: "If we want to take this conversation even further into the economics of music streaming . . . globally streaming grew 9.4 percent in the last year. Music streaming today contributes to 43 percent of music revenue. On a broader level, there are 1.6 billion people who stream music."[11] That the rhetoric of data—or what some would call "dataism"—would apply to mainstream music and music platforms is to be expected; after all, the world of pop and major labels is driven by numbers (plays, views, hit charts, sales, fan base stats, and so forth). When imaginations of the future for indie music practices are primarily enabled by data, numbers, trends, algorithms, and the affordances and governances of platforms, it is alarming.

I read a lot of numbers in this volume. Relatedly, I came across extensive discussions on the algorithmic, infrastructural, and technological intricacies involved. They are essential. As Campos Valverde notes, "Despite the prevalence of music streaming for the past fifteen years, we still know very little about the design and architecture of its software infrastructures and their impact on society. . . . The role of specific technologies, and the political-economic forces shaping them, has been insufficiently explored."[12] At the same time, while acknowledging the importance of the technological, one must also remain alert and refrain from reiterating its importance. *Technology* and *culture* are two keywords in Hesmondhalgh's introductory chapter. As I make my way through the chapters of this volume, I sometimes wonder whether I am reading about MSP technology or culture. A few years ago, some colleagues and I coauthored an introductory piece to a special issue on platformization of Chinese society. There, we made an appeal: "The study of platformization needs to be opened to the realms of the social and the cultural."[13] In proposing creative ways to disturb the algorithmic, Sophie Bishop and Tanya Kant urge that "researchers can better discern how technology users make sense of their data, the ways in which identity can be co-constructed by social media platforms, and how our interactions with technology ultimately shape social lives in meaningful and highly affective ways."[14]

The users. My appeal would be broader: to rally the personal and disturb the algorithmic, data, numbers with the personal. Perhaps more cultural studies, less platform studies. Or more cultural studies in platform studies. In Tiziano Bonini and Emiliano Treré's formulation, "platforms are a battleground where people sometimes dance with algorithms and other times clash with them. Sometimes they lose; other times, they (temporarily) win. Sometimes they game the system; sometimes they radically change it."[15] While I underwrite their inquiry surrounding algorithmic resistance, agency, and politics, my remit here is to be inspired. In the following, I am suggesting ways, inspired by the chapters, in which we could write from a more personal point of view, incorporating more of the personal into MSP technology as a tactic, a means of resistance, and a way of engaging with more stories, better stories.

MORE PERSONS, FEWER NUMBERS

In their chapter in this volume, Gómez, Gallego, and Muñoz-Larroa present a fascinating case study of corridos tumbados, "a contemporary subgenre [of regional Mexican music] that merges traditional corridos (narrative ballads) with elements of trap and hip hop." Tracking the success story of corridos tumbados and

Mexican music at large—success as measured by their performance on MSPs in the Latin American context—the authors witness "new patterns of music circulation via streaming that favor countries and artists who previously lacked such visibility and reach," writ large in the increasing popularity of corridos tumbados.[16] The authors offer some factors contributing to this Global South success story—musical, cultural, and industrial—complemented by views from the frontline creative workers: independent producers, musicians, and managers.

This is a robust way of mapping what the authors call "the new dynamics of the music industry in the streaming age." At the same time, I start to wonder: if we do not follow a successful genre but a successful artist, what might we harvest? To push it even more, what if we leave behind the label "successful"? That is to say, if we are guided not only by numbers, but by people, would that be another mode of knowledge production on the impact of MSPs? What I am suggesting, I gather, may not be very different from ignoring the musical options recommended to me by the algorithmic systems and choosing for myself. Some artists working in corridos tumbados are mentioned in the chapter. I am curious about how they achieve such algorithmic visibility, as much as I am curious about the "failure" stories of those—many, no doubt—who have not made it or do not care to make it.

Elsewhere, I have proposed a method: "follow the person."[17] I have taken my cues from Scott Lash and Celia Lury's method of "following the object."[18] Thinking along the fault lines between classical cultural industries and global cultural industries, Lash and Lury's central concern is to go beyond (media) representation and foreground movement—that is, the dynamics, entanglement, and interplay among agents in cultural production. I share the concern and, in this case, their method. To follow the object, for Lash and Lury, is "to consider the markets of the global culture industry as neither pre-given nor static, as neither simply global nor as a merely local, but as dynamically constituted by the movements, the biographies, of objects."[19] We can, and should, do the same with a person: write more stories in our streaming age from a more personal starting point. Sometimes, if I do not heed the recommendations on Spotify, I may come across someone, something, some story—unexpectedly, contingently, probably with more bewilderment and provocation.

EXCEPTION AS METHOD

In some sense, this is an oblique response to the rejoinder voiced by feminist and queer data studies. What they call for is a critical approach toward data practice that is sensitized by the concept of intersectionality and the ideals of diversity, inclusiveness, and equality. They draw researchers' attention to more traditional methodologies that tend to erase or marginalize certain voices and experiences, such as those of women and queer populations. In her powerful analysis of the algorithm as oppression—mechanisms that reinforce and sustain racist and sexist practices—Safiya Umoja Noble makes a dazzling move by calling forth the algorithm as an ally. Her mandate is to show how algorithms can transform them-selves into agents for greater social justice and a better world.[20] While Noble is discussing primarily race and gender politics in her book, her argument seems equally pertinent to other axes of inequality, such as North-South. As Noble cautions at the very outset, "Part of the challenge of understanding algorithmic oppression is to understand that mathematical formulations to drive automated decisions are made by human beings."[21]

From this cautionary note and the general attention needed for to address mar-ginalization, I continue my "follow the person" appeal. There is a need to learn from those working for MSPs (as Sprengel did with people working for the plat-form Anghami). My appeal is more general. In the section above, I mentioned that I was curious about "success" and "failure" stories. Here, I want to challenge the tyranny of numbers by proposing that we listen not only to those working in the algorithmic industry but also to those in the music industry who do not command numbers, big data, or algorithmic attention. As mentioned above, it may be a matter of not making it,[22] or a choice not to make it—some music prac-titioners may prefer to keep it small, niche, and free. It is often not only a personal issue; structures may dictate that some artists and genres remain on the periphery. As a lyricist born and bred in Hong Kong, I am drawn to how minority language (Cantonese, as opposed to Putonghua) and geopolitical-cultural periphery (a southern city, as compared to, say, Beijing, the capital) affect local music's develop-ment, whether on MSPs or not. We should rally algorithms as our allies, but some-times we may need to ask human beings to be allies too. The aspirations, dreams, pleasures, pains, struggles, and everyday lives of music practitioners do not neces-sarily speak the language of numbers or capital, but they may tell us things that algorithms tend to silence. If I may detect methodologies in this book generally determined by numbers, I propose employing exceptional stories (by definition, different, and thus more numerous than the master narrative) recounted by music practitioners (not the successful, mainstream-centric ones) to understand what is happening: exception as method.

THE STATE, THE COPYRIGHTS . . .

In Onur Sesigür's chapter, we hear interviews with ten musicians: "None felt it was possible to earn enough from streaming to pay their bills, at least in their current situation." Admitting the impact of MSPs, they, however, tease out the intricacies of that impact. Streaming serves not as a major source of income but as promo-tion for live gigs and concerts, which have become "the main source of income for musicians who choose to bet on their own music, with streaming serving as an agent of exposure." These Istanbul-based music practitioners add that this scenario is applicable not only to them and their peers but superstars too. While I plead for the personal, I do not disregard numbers—this set of figures cited by

Sesigür is telling: "For a musician to cover the monthly cost of living in Istanbul (US\$617.30), they would need roughly 1.5 million streams. Meanwhile, a musician in Reykjavik only needs around two hundred thousand streams (to cover the average rent there of US\$1307.70)." Iceland's pay-per-stream (PPS) rate stands as the world's highest—US\$0.0067, compared to Turkey's at US\$0.0004.[23]

I am touched by this account; the predicament resonates with the experiences of many music practitioners in Hong Kong, with whom I am more familiar. If asked, they would likely say something similar: the songs they create serve primarily as resources for front-stage artists to gain airplay, views, and streams—essentially marketing tools to secure product endorsement contracts and concert ticket sales, where the real money is. And this is not an issue specific to streaming: this uneven distribution of music-generation revenue emerged when "normal" music sales and airplay revenue dropped dramatically. As a lyricist, I receive royalties from these two sources, but not from the artist management side of the music business. Why not? We miss our fair share under the current—perhaps overdue—copyright regime, even though my fellow songwriters and lyricists and I have been supplying the music, the very resource that makes an artist a star, and generates economic value. Many contributions in this book recount a similar phase in pop music history: the emergence of digital reproduction technology and piracy, which infringed on copyrights and led to the current juncture, where streaming technology has effectively rendered piracy obsolete and the copyright regime, as we know it, has resumed.

Such a missed opportunity to reflect and rebuild—I sighed when I was reading about this phase of piracy and the near-collapse of music copyright regimes. Following the personal and listening more to music practitioners may tell us more stories about the complex interaction of technology and culture, yielding new questions that interrogate how capital works. Quite apart from the disparity in PPS earnings from copyright between the Global North and South, music practitioners may ask a more fundamental question about the distribution of music-generation wealth, moving beyond the Western-led discussion on copyright and intellectual property.[24]

And we are yet to cover more than the economic. Here I turn to another contribution: Zhongwei (Mabu) Li and D. Bondy Valdovinos Kaye's inquiry into one Chinese case. Li and Kaye conducted focus group discussions with independent music creators about their experiences self-releasing on MSPs. One of the consequences of the institutionalization and conglomerization of MSPs is the facilitation of top-down censorship, plus the platform's own practice of self-censorship. According to them, censorship on MSPs is more common and stricter than on the Chinese internet. Li and Kaye's contribution is to track the history of platformization in China and map out various dimensions of these creators' engagement with MSPs. How I wish Li and Kaye could report on their discussions with these creators about their survival tactics—not only economic, but also political. The state is often missing when we examine the numbers and the capitalist logic and practices driving MSPs. In many localities in both the Global South and Global North, the state—or often the state-market nexus—is highly present, with surveillance and censorship more often the rule than the exception. In 2019, the music of an electronic duo I have collaborated with was removed from all mainland Chinese MSPs, as well as Apple Music, for reasons generally known to be political.[25] Sesigür, Li, and Kaye are not the only authors who have talked to music practitioners. The point I am making by way of their work is that a more personal approach may tell more stories, encapsulating a different set of problematics and questions—such as those on copyrights and the state—surrounding MSPs.

GENDER, AGE . . .

Put another way, this is to recuperate the geopolitical as embodied in the personal. In an article on creative class mobility, I highlight the importance of including the geopolitical, urging that we pay particular attention to local configurations and related power dynamics—such as those between the periphery and the center—when examining why and how creative workers move, specifically in the context of Hong Kong and mainland China.[26] I believe it is relevant to research on MSPs. But then the geopolitical is merely one dimension of the personal. In the same article, I posit the intersectional, particularly the need to insert a gendered perspective. From the narratives I gathered from creative workers, it was evident how female practitioners experience the industry differentially and inequitably. I thus reiterate the call from a line of extant scholarship to engage gender relations and subjectivities more consciously and explicitly when researching creative practitioners.[27] A recently published edited volume on streaming platforms and Indian cinema has taken up gender as its linchpin, which underwrites its importance as well as its persistent elision from research.[28]

While the aforementioned volume dissects gender politics in terms of representation, I must stress its relevance for production issues, which reminds me of Hesmondhalgh's call to turn toward production in cultural industries.[29] In their chapter on the Mexican experience, Gómez, Gallego, and Muñoz-Larroa discuss at length the rising popularity of corridos tumbados, in connection with MSPs. While the authors attribute their popularity to the affordances of digital platforms, which help "young northern Mexicans" express and disseminate their music, they also mention the controversy caused by the genre's alleged lyrical glorification of drug use, weapons, and materialistic excesses. What caught my attention is that among the list of corridos tumbados artists who have made headway on MSPs, "males predominate."[30] This passing remark deserves more attention, research, discussion, and intervention. How would a gendered perspective yield different and more stories and knowledge about those working in and with MSPs? This question lingered as I read the chapters of this book.

It is not only gender. The intersectional approach reminds me that when I follow anyone, I must stay sensitive to "the intersection of multiple categories—and proceed to uncover the differences and complexities of experience embodied in that location."[31] In other words, to have more stories—and more personal stories—surrounding MSPs or music-making in general, we need to inquire not only in terms of their professional category but also in relation to other categories: geopolitical, gender, age, and so forth. If our concern is equality, the examination and discussion should not be confined to a North-South divide, but rather extended to as many other domains as possible. I name age here, in response to the "young northern Mexicans" quoted above. Indeed, to pose the question differently, how would an age perspective yield different and more stories and knowledge concerning those working in and with MSPs? Especially in the so-called streaming age—in which technological savviness is a major prerequisite for engagement and, to use the *s*-word again, success—an inquiry into MSPs in connection with different age cohorts, and perhaps aging in general (particularly, how to keep up with presumably fast-paced technological developments), would be illuminating.

THE USERS, THE USES . . .

It is equally illuminating to note that the research I found on the connection between music streaming and older populations shows certain features. One is to see how savvy they are in using streaming technology, often viewed as an extension of radio; the other is to see older people as users, with the older generation of radio listeners now having switched to MSPs.[32] This is necessary research, but what I suggested above refers to something else. My curiosity goes to the various possible relationships between different age cohorts and streaming technology; one may well be participatory, such as older people making music and posting on MSPs. That said, I want to revert to the very issue of users. According to Mao Mao and David Good's study of people over forty in the United Kingdom, sociality emerges as the most important motivation for older people's use of MSPs. In particular, their membership in social (music) groups is a good indicator of their willingness to engage with streaming technology, such as heavier use of YouTube or Spotify; they want to stay tuned and share.[33]

What about other groups of MSP users? Here I reiterate what I cited above—the appeal by Bishop and Kant to researchers of technology: how users make sense of data. They highlight the possibilities of identification, meaning-making, and the affective.[34] In Manabe's account of streaming in Japan, learning from users—or listeners—serves as an important thread, even when the account is presented without the conventional method of audience research. "The charts did not reflect actual listening habits," she states, after discussing chart performances and CD sales in Japan. This statement is a clear reminder of the indexical value of numbers regarding what people do with music. Later in the chapter, Manabe summarizes an interview she conducted with Ono Tetsutarō, who discusses the way streaming technology enables the discovery and appreciation of older music, noting, "The top-streamed songs are often those that have been out for some time, as people listen to a song repeatedly over time."[35]

I want to know more about users than the contributors to this volume have been able to offer. For example, how do users start this mode of streaming? Why? With whom? On which platforms? What are the particular features that attract them? What kind of older music do listeners play in different nations and regions? In this regard, I am also cued to a local experience of mine. Above, I mentioned an electronic duo whose music was banned on mainland Chinese MSPs—as well as Apple Music, I must add. Occasionally, however, I would come across Chinese listeners who tell me the following. They are listening to some of the newer songs I have just released and find them interesting. They search the internet and come across titles of the works I created with the duo but cannot find them on any MSPs. Further online searching would inform them of what has roughly happened, and sometimes they would still be able to access those songs via other means—online and/or offline, sometimes thanks to a VPN. There are many instances like this— emerging in the warp and weft of online and offline practices, the personal and political, surveillance and resilience, technology and culture—toward which users would guide us; not the number of airplays or views, not hits, but the habits, with all the meaning-making and life-making processes involved, that tell us what is going on.

CODA: THROW SOME STONES

In the chapter on the Japanese history of music streaming, Noriko Manabe refers to Eric Drott's book *Streaming Music, Streaming Capital*. Specifically, Manabe points to the metaphor of streaming that "likens music to water, betraying an ideology that devalues it as a kind of utility—a ubiquitous background that is not fully appreciated."[36] I too am drawn to Drott's critique of the streaming metaphor: by likening music to water, it naturalizes certain attributes of music and music streaming. As formulated by Drott, "To the extent the 'music like water' metaphor remakes music in the image of water, music too begins to resemble something simply given, as just another feature of the ambient cultural environment, whose readiness-to-hand for listeners occludes the work of the musicians who produced it."[37]

It is at this juncture that I return to the technological determinism I mentioned at the beginning of this afterword. Sally Wyatt reminds us to treat technological determinism seriously, just as we should treat technology itself seriously.[38] The contributions to this volume form a testament to the benefits of treating technology seriously, and taken together, they form a robust rebuke to technological determinism. Wyatt understands technological determinism as "a broader public discourse which seeks to render technology opaque and beyond political intervention and control," and part of this discursive formation lies in the use of metaphor.[39] As Wyatt reminds critical scholars working on internet and digital media, we must bear in mind Donald McCloskey's warning that "unexamined metaphor is a substitute for thinking—which is a recommendation to examine the metaphors, not to attempt the impossible by banishing them."[40]

We need to understand the genealogy of metaphors,[41] locate misleading ones, and consider the power of our own words and metaphors.[42] This is Wyatt's appeal to us, and I consider the chapters in this volume, in their different ways, to be an intervention through words and metaphors, or an interrogation of the master metaphor of streaming. I will end with a metaphor—or, more precisely, a metaphorical saying, a Chinese one, for that matter: *one stone stirs up a thousand waves.* The way I read this saying and appropriate it for our purposes is this: It requires a person to throw the stone, to disturb the flow, to denaturalize the stream and streaming, to stir up and evoke such rippling effects that must be left to their own devices—undecided, even unintended, and yet consequential. This is the personal, contingent, and affective side of things that I mentioned above. Let us continue throwing stones. The more, the better.

NOTES

1. David Hesmondhalgh, "Streaming's Effects on Music Culture: Old Anxieties and New Simplifications," *Cultural Sociology* 16, no. 1 (2022): 4.

2. An exception is a piece I coauthored to introduce a special issue on platformization and China. See Jeroen de Kloet, Thomas Poell, Zeng Guohua, and Chow Yiu Fai, "The Platformization of Chinese Society: Infrastructure, Governance, and Practice," *Chinese Journal of Communication* 12, no. 3 (2019): 249–56.

3. Lawrence Grossberg, "On the Political Responsibilities of Cultural Studies," *Inter-Asia Cultural Studies* 11, no. 2 (2010): 241–47.

4. See for instance Megan M. Wood, "On 'Telling Better Stories,'" *Cultural Studies* 33, no. 1 (2018): 19–28.

5. Sally Wyatt, "Technological Determinism Is Dead; Long Live Technological Determinism," *The Handbook of Science and Technology Studies* 3 (2008): 165–80.

6. Marshall McLuhan, *Understanding Media: The Extensions of Man* (1964; repr., Cambridge, MA: MIT Press, 1994).

7. See for instance Yuval Noah Harari, "Dataism Is Our New God," *New Perspectives Quarterly* 34, no. 2 (2017): 36–43.

8. From Hesmondhalgh's chapter in this volume (chapter 1).

9. Yiu Fai Chow, Jeroen de Kloet, and Leonie Schmidt, *It's My Party: Tat Ming Pair and the Postcolonial Politics of Popular Music in Hong Kong* (London: Palgrave Macmillan, 2024), 4.

10. From Manabe's chapter in this volume (chapter 7).

11. From Sprengel's chapter in this volume (chapter 5).

12. From Campos Valverde's chapter in this volume (chapter 12).

13. De Kloet, Poell, Guohua, and Fai, "Platformization of Chinese Society," 249.

14. Sophie Bishop and Tanya Kant, "Algorithmic Autobiographies and Fictions: A Digital Method," *Sociological Review* 71, no. 5 (2023): 1013.

15. Tiziano Bonini and Emiliano Treré, *Algorithms of Resistance: The Everyday Fight against Platform Power* (Cambridge, MA: MIT Press, 2024), 178.

16. From Gómez, Gallego, and Muñoz-Larroa's chapter in this volume (chapter 6).

17. Yiu Fai Chow, "Hope against Hopes: Diana Zhu and the Transnational Politics of Chinese Popular Music," *Cultural Studies* 25, no. 6 (2011): 783–808.

18. Scott Lash and Celia Lury, *Global Culture Industry: The Mediation of Things* (Cambridge: Polity Press, 2007).

19. Lash and Lury, *Global Culture Industry*, 18–19.

20. Safiya Umoja Noble, *Algorithms of Oppression: How Search Engines Reinforce Racism* (New York: NYU Press, 2018).

21. Noble, *Algorithms of Oppression*, 1.

22. See Sesigür's troubling account in this volume (chapter 11) of how streaming fails to generate enough income for music practitioners in Istanbul, who then resort to creating commercial jingles.

23. See Sesigür's chapter in this volume (chapter 11).

24. For a related discussion on "property," see Hendrik Theine and Sebastian Sevignani, introduction to "Media Transformation and the Challenge of Property," special issue, *European Journal of Communication* 39, no. 5 (2024): 407–11.

25. Chow, de Kloet, and Schmidt, *It's My Party*.

26. Yiu Fai Chow, "Exploring Creative Class Mobility: Hong Kong Creative Workers in Shanghai and Beijing," *Eurasian Geography and Economics* 58, no. 4 (2017): 361–85.

27. See for instance Brooke Erin Duffy, "The Romance of Work: Gender and Aspirational Labour in the Digital Culture Industries," *International Journal of Cultural Studies* 19, no. 4 (2016): 441–57; and Angela McRobbie, *Be Creative: Making a Living in the New Culture Industries* (Hoboken, NJ: Wiley-Blackwell, 2016).

28. Runa Chakraborty Paunksnis and Šarūnas Paunksnis, eds., *Gender, Cinema, Streaming Platforms: Shifting Frames in Neoliberal India* (London: Palgrave Macmillan, 2023).

29. David Hesmondhalgh, *The Cultural Industries*, 4th ed. (London: Sage, 2019).

30. See Gómez, Gallego, and Muñoz-Larroa's chapter in this volume (chapter 6).

31. Leslie McCall, "The Complexity of Intersectionality," *Signs* 30, no. 3 (2005): 1782.

32. See for instance Hugh Chignell and Kathryn McDonald, eds., *The Bloomsbury Handbook of Radio* (London: Bloomsbury Press, 2023); and Mao Mao and David A. Good, "Understanding the Use and Motivation of Digital Music Technologies among Middle-Aged and Older Adults," in *HCI '18: Proceedings of the 32nd International BCS Human Computer Interaction Conference* (Swindon: BCS Learning and Development, 2018), http://dx.doi.org/10.14236/ewic/HCI2018.47.

33. Mao and Good, "Understanding the Use and Motivation."

34. Bishop and Kant, "Algorithmic Autobiographies."

35. See Manabe's chapter in this volume (chapter 7).

36. Eric Drott, *Streaming Music, Streaming Capital* (Durham, NC: Duke University Press, 2024), cited in Manabe's chapter in this volume (chapter 7).

37. Drott, *Streaming Music*, 96.

38. Wyatt, "Technological Determinism," 175.

39. Wyatt, "Technological Determinism," 176.

40. Donald N. McCloskey, "The Rhetoric of Economics," *Journal of Economic Literature* 21, no. 2 (1983): 481, cited in Sally Wyatt, "Metaphors in Critical Internet and Digital Media Studies," *New Media and Society* 23, no. 2 (2021): 406–16.

41. See for instance Maureen McNeil, Michael Arribas-Ayllon, Joan Haran, Adrian Mackenzie, and Richard Tutton, "Conceptualizing Imaginaries of Science, Technology, and Society," in *The Handbook of Science and Technology Studies*, ed. Ulrike Felt, Rayvon Fouché, Clark A. Miller, and Laurel Smith-Doerr, 4th ed. (Cambridge, MA: MIT Press, 2016), 435.

42. Wyatt, "Metaphors," 412.

Index

9 780520 409057